How Space and Motion Determine Biological Form

How Space and Motion Determine Biological Form

A Structural Approach to Development and Evolution

Tim Haug

Copyright © Tim Haug

All Rights Reserved

ISBN - 9798335415316

TABLE OF CONTENTS

Preface

1. Introduction   1
    The Elements of Form

2. Form   9
    Simple and Complex Organization
    Categories of Structure:
    Levels of Organization
    Units of Form
    Variety

3. Space: Environment and Position   31
    Environment
    Position
    The Conversion of Matter

4. Motion: The Dynamics of Form   46
    Circulation and Flow
    The Biological Reaction
    The Law of Conservation of Mass
    Growth and Maturity
    Reproduction
    Phyletic Size Increase and Decrease
    Timing of Developmental Events
    Form as a Pattern of Motion

5. Regulating Motion   71
    Feedback loops
    Efficiency
        Molecular Efficiency
        Increasing Efficiency
        Efficient Movement of a Population
        Efficient Resources
    Homeostasis
    Communication

6. Determinants of Differentiated Form   95
    Variation and Adaptation
    Molecular Synthesis and Differentiation
        Molecular Organization
        Protein Synthesis
        Genetic Replication and Expression
        Gene Position
        Summary
    Cell Differentiation
        Embryonic Cell Differentiation
    Caste Differentiation
    Summary

7. Diversification and Stasis   119
    Characteristics of Development
        and Maturity Periods
    Anagenesis and Cladogenesis
    Diversification
    Sequential Development
        Protein Folding Sequence
    Karyotype Evolution
    Stasis

Afterword   141

Glossary   147

PREFACE

Structural evolution is the idea that all biological form can be reduced to a few categories of structure that are common to all organisms and that nature, following a few basic rules, organizes these structures into the myriads of organisms that have existed. It is based on the premise that a population expands only to the limit of its environment. This is a consequence of Liebig's law, and while this premise is mundanely recognized throughout nature, it has profound implications for development and evolution that are hardly self-evident. However, within the conceptual framework of structural evolution in which biological form is reduced to categories of common structures, Liebig's law plays an essential role in the organization of biological form.

A structural approach to biological organization provides a context in which to describe how nature organizes units of form into categories of structure. In this perspective there are levels of organization that are constructed in the same way by the same processes, and they form a hierarchy of biological organization. It supposes that each form is assembled individually through a process of development that follows certain physical rules of motion and that motion is limited by an individual's ability to move and the space in which it moves. Instead of examining the phyletic history of a species to determine its path of evolution, a structural approach considers the things are necessary to construct an organism and how those things come into being and come to be assembled as an organism.

The hereditary system of organization established by cladistics simply does not provide the context in which this interpretation can be made. It presents no scheme of structural organization and no basis for assuming any particular unit of form. Development and evolution are seen as distinct processes, and evolution is directed by the social interaction of competition. In contrast, structural evolution sees development and evolution as the same process carried out at different levels of organization and directed by the same rules of physical organization. More generally, the hereditary view considers form to be a fixed entity unless genetically modified, while the structural view considers form to be a fluid pattern of motion in which an organism is the pattern of motion of its molecules and cells, and its lifespan is merely the duration to which that pattern persists.

Organizing biology into structural categories is rather simple in comparison to hereditary classification because there are only a few categories to contend with, instead of the ever-expanding multitude of species that are the Linnaean system. This does not make it a better system, it simply makes it a different system, and because it is different, it tells us very different things about biology than does cladistics. It draws our attention to insights that cladistics misses by not providing the context in which to draw the connections between seemingly disparate processes. Conversely, the context of structural evolution provides very little, if any, insight into the hereditary relationships of organisms.

Cladistics is in no way inferior to a structural approach; it tells us a vast amount about biology and biological history that a structural approach cannot. However, the same is true for a structural approach, it tells us much that a hereditary approach cannot. The two views are simply separate ways of interpreting biology. They are not contradictory, but instead rather complimentary to each other in providing a wholistic picture of biological organization. They are different views that focus on different aspects and elicit different ideas about same phenomena.

The initial chapters on form, space, and motion set out the foundational concepts of a structural approach to biological organization by identifying the structures common to biological organization and describing how they relate to one another. This requires redefining many terms commonly used in the parlance of biology, as well as adding a few new ones. Such terms as individual, population, system, unit of form, environment, position, and complexity, have meanings specific to this context. Other terms such as circulation, flow, conversion, and efficiency are added to build out a broader concept of biological organization. This sets the stage for the final chapters in which the concepts and definitions established in former chapters are used to describe the developmental processes of diversification and stasis.

Many new concepts, found in no other discussion on biology, are presented throughout this discourse. Elements of form, categories of structure, levels of organization, direct and indirect resources, to name a few. These not only help to build the conceptual framework of structural evolution but also lend explanation to many of the current problems of development and evolution. Although none of these concepts are particularly enigmatic, and many are co-opted from similar ideas found elsewhere in biology, most of them are more specific than in their common usage, and some are entirely new to biology. However, a clear understanding of these concepts is necessary to grasp the full meaning of what structural evolution is. For that purpose, definitions for these terms, when different from the common usage, have been provided both in the text and the glossary.

Whether or not one accepts the validity of the concepts as presented is mostly left for the reader to decide. Aside from some anecdotal evidence, little validation of these ideas is provided. However, I have tried to present the reasoning that underlies them. To substantiate every concept, idea, assertion, or claim made in this book with material and experimental evidence would have prevented its completion, and in some instances, there is yet no data that either supports or refutes the concept. Moreover, what is presented here is merely a different way of interpreting the facts and data that exist, and which are commonly known to biologists. In my view, adding substantial supporting data seems more likely to obscure the presentation rather than enhance its interpretation.

1. INTRODUCTION

Throughout the latter half of the twentieth century and into the twenty-first, the field of evolution has focused on genetics. It is generally understood that our genes determine who we are. They are thought to determine what species we are, how we develop, our shape, our physical characteristics, our abilities, what diseases we may suffer or be immune from. They may even determine how long we live and affect what we do while we live.

We understand how genes are transcribed into RNA, how transcribed genes build protein molecules and that protein molecules are the basic building blocks of every organism. Geneticists can even map an organism's entire genome. Much is known about how individual genes function, and more is discovered each year,

But much is also not known. A genome seems to know when, where, and how many of each type of protein to build, it also knows the specific kinds of proteins that are needed for each of the various kinds of cells that make up an organism, and it appears to know even more than this. It knows when to build different types of cells, where to build them, and how many of each are necessary to construct its specific organism. This is an enormous amount of information supposedly inscribed into the genome. The genome constructs an organism's anatomy, molecule by molecule, and yet that anatomy appears to be encoded nowhere in the genome's structure.

Geneticists are now beginning to realize that a genome is not a set of instructions that direct the formation of an organism but rather only a list of the parts necessary to build the organism. Sequencing a genome is much like producing a list of words or a dictionary that lists only the words necessary to write a particular story. Many of its words are used in most of the stories we know, and different stories can be written with the same words, but many other stories need additional words. We can write a dictionary for every story we know, but if the dictionary is not labeled, we would not know which story it is for. Most importantly, we understand how words are arranged to form sentences, paragraphs, and chapters, but we do not know how the dictionary for a particular story knows to arrange the words it contains into its story.

The research conducted since DNA was first discovered in 1869 by Friedrich Miescher, has produced an incredible knowledge of the structure and function of DNA, but we still do not understand how the complement of genes that comprise a genome knows to build a particular organism. We know how a protein is constructed and how a genome builds a specific protein, but how does the genome know to build the specific array of proteins necessary to construct a certain kind of cell, or the array of cells necessary to build a particular species of organism?

Although chromosomal structure – the karyotype – is specific to the species, there is nothing contained within the DNA that distinguishes the allometry of a species. In other words, one could not deduce a species from its genes, or conversely, deduce chromosomal structure from the species, without prior knowledge that the two are associated. The one important question left unanswered by genetic research is where and how is an organism's anatomy encoded in its DNA.

In this book we are going to look at evolution and development as if we were to set about constructing an organism from scratch brick by brick, assembling molecules into cells, cells into organisms and organisms into ecosystems. What information is needed to do this and how does nature acquire and use this information to produce an organism?

Imagine a home builder being given a list that specifies all the various materials needed to build a house, but the list does not specify the quantity of each kind of material needed, nor are there plans that tell where in the structure of the house different materials should be placed. It is just a list of the kinds of parts needed to construct a house. Many, if not a majority, of these parts are used in most varieties of houses and often in many different places in the same house for different purposes. How does the builder know what kind of house is specified by the materials list he is given? If an organism is the house and nature the builder, how does nature know to build the proper house?

Much like the materials list a builder uses, the chromosomes are a list of genes needed to construct an organism, the genes specify proteins, but how many of each and where they belong in the organism's structure is not specified. A builder no doubt knows how to assemble the parts of house, but without plans or instructions he does not know how many of each kind of part he needs nor where they should go in the structure; and having only a list of parts gives no indication what the final home should look like.

Genetics presents a similar problem, but with an additional degree of difficulty. The genome contains a list of genes, but organisms are not built with genes, they are built with proteins and genes do not necessarily specify specific proteins, they specify a string of nucleotides. These are protein precursors, short stretches of DNA, that by themselves are not functional. But when genes are transcribed into RNA molecules, and RNA molecules are translated into polypeptides, and when various polypeptides, often produced from different genes, are strung together, then they can become a functional protein. And a functional protein is a useful building block in the construction of a cell.

So, in this construction analogy, it is as if the genome's list of parts given to nature's contractor does not even identify the building blocks necessary to build the organism, but rather lists those things necessary to make the building blocks used to build the organism. It is as if our builder's materials list identifies the clay, lime and sand that go into making concrete, but does not specify the ratio in which these substrates need to be mixed. Nor does it identify whether that concrete is to be formed into bricks, blocks, or tiles; nor whether those bricks, blocks and tiles are to be assembled into walls, foundations, floors or rooves.

Astoundingly, nature uses just such an abbreviated list found in the DNA to build bones, exoskeletons, hearts, nervous systems, and everything else an organism may need to be functional in its environment, and it does so over and again with immense accuracy. How does a gene on such a list know when to replicate itself so that it makes the correct protein needed for a particular type of cell that is needed in a specific organ of an organism? Moreover, that same gene may, at some other later time during development, be used to build a slightly different protein for a different cell in a different part of the body. How does it know to do that?

It is not enough that DNA has codons to tell it when to start and stop transcribing a particular gene. It must also know when that gene is needed and how many times it must be transcribed to produce the necessary quantity of proteins. To do that, the gene must know what kind of cell it wants to build, and for that, it must know where the cell is within the organism. Codons seem only to identify a section of the DNA that constitutes a particular gene and nothing more.

For a properly functioning genome to replicate the correct gene, at the correct time and place, it must in some way receive information that tells it which gene is needed, when it is needed, and how often to transcribe it. This is especially important in embryonic development. During this phase, the genome needs to know its location within the developing embryo so that it knows what kind of cell to build and what proteins are needed to build that type of cell. Coincidentally, embryonic development is a time when the location of a cell within the organism is most fluid. A differentiating cell in an embryo not only needs to know where it is, but also where within the organism it will eventually be.

Each cell within an organism has the exact same set of genes, but at different times and places within the organism these identical genomes function differently. Yet, they always function in concert with one another. Especially, and most remarkably, during development. For this to occur, there must be a means through which information is transmitted among the cellular genomes of an organism. That information tells the genome which, when, where, and how often a gene should be transcribed.

Interestingly, this information can also be shared among unrelated genomes. This feat of nature occurs when two distinct genomes work together to build a chimeric organism. A chimera is an organism in which two embryos fuse together in the initial stage of embryonic development. The fusion of embryos causes the resulting single embryo to have two distinct genomes. Each genome, although programmed to build an entire organism, builds only that part of the organism complementary to the part built by the other genome. The result is a single organism carrying two distinct sets of chromosomes. How does one genome know to build only those parts of an organism that the other genome does not build? The information necessary to do this must in some way be imparted to both genomes.

Although much is known about how genes function, how they are transcribed and translated into polypeptide strings, and how they eventually become proteins, we know very little about how a particular genome, residing within a cell, within an organism, directs the organism's development. How do genes know which kinds of proteins are needed to make a particular type of cell? How do the genes in one part of the body know to differentiate into one kind of cell, while the same genes in another part of the body know to differentiate into another kind? How do genes tell cells when to start and stop differentiating? How do genes know when enough cells, of a certain type necessary to produce a functioning organ, have been produced? How do genes know what the shape of that organ should be? These are the questions that will be considered throughout this book.

THE ELEMENTS OF FORM

Whenever humans go about constructing something, there are four variables to contend with: variety, volume, timing, and location. Whether we are building a home, a car, a bicycle, or even painting a picture, we need to have a variety of parts. For a picture, the variety of parts are the different colors of paint. We also need to have those parts available in a volume or number necessary to complete our project. Without sufficient screws and bolts we may be unable to assemble a functioning bicycle, or without sufficient paint we cannot paint the entire picture. We will want to assemble those parts in an order that is both efficient and allows us to completely assemble the project. For example, it is easiest to construct a house by laying the foundation first rather than last, and certain phases of the plumbing need to be added at specific times during the home's construction. Finally, all the parts need to be assembled in the correct place. If the wheels of a car are attached to the roof rather than the axle the car will not function properly.

These same variables of location, timing, quantity, and variety must be controlled during an organism's development. So, whether considering a protein, a cell, or an organism, the key to why a form develops as it does, is in knowing why a particular variety of parts is produced, in the number they are produced, and at the time and place they are produced. Those parts may be the nucleotides that form a protein, the proteins that form a cell, the cells that form an organism, or even the organisms that form an ecosystem. If we can discern those factors that control these variables during development, we may then understand why form develops as it does and why it evolves when it does.

Presently, form is assumed to be determined by heredity. Its development is controlled by the genome and variation in the genome is caused by a substitution of genes arising from random mutations that become fixed through competition. Each form is the end-product of its genetic evolutionary history.

In contrast to this heredity view of development, a structural view supposes that each form is assembled individually through a process of development that follows certain physical rules of motion and that the movement of the compositional parts that construct a form is limited by their ability to move and the space in which they move. Instead of examining the phyletic history of a species to determine its path of evolution, a structural approach considers what things are necessary to construct an organism and how those things come into being and come to be assembled into the organism.

At the heart of this structural approach is the scientific method which assumes a process repeated will produce the same result. Applied to biology, this assumption presumes that to the extent a developmental process is repeated, the form it produces should be replicated. Variations in form may then result, not only from a substitution of compositional parts, but also from changes in the process that alter the timing of events, the location where events occur, and the number of the various compositional parts produced. This is quite different from the heredity view which assumes that all variations ultimately arise from substitutions at the molecular level of the gene; and supposes no physical rules of organization.

In the following pages we will explore the concepts of structural organization by describing how the four elements of form – variety, location, number, and timing – are determined by physical and biological processes. In this description, development and evolution are the same process operating at different levels of organization. Levels of organization are created by three specific categories of biological structure, and those categories of structure are individuals, populations, and systems. These structures are organized according to physical laws of motion that govern their interaction and produce causal relationships among environmental space, motion, and biological form. In this view biological form is dynamic with its compositional parts moving in continual patterns of motion. Variation, adaptation, diversification, stasis, development, and evolution are all consequences of this movement and the space in which that movement is confined.

2. FORM

SIMPLE & COMPLEX ORGANIZATION

Both physical and biological organization produce a myriad of structures from the very small to the very large. In the physical realm, structures are categorized into groups of similar structures based on three things: the similarity of their motion, the similarity of their substructure and how they fit into a superstructure. This, for example, is how we identify a solar system, by how it fits into the superstructure of a galaxy, by how it moves within a galaxy, and by its planetary system substructure. This is also how we innately understand things to be more or less complex, by their motion, their substructure, and their superstructure. A solar system, for example, is less complex than a galaxy but more complex than a planet. We know this because planets must be organized to create a solar system, but to create a galaxy, planets must be organized into solar systems and then solar systems must be organized into a galaxy. It is one step to organize planets into a solar system and an additional step to organize solar systems into a galaxy. This additional organizing step is additional complexity. Of course, a galaxy does not develop in this step-by-step fashion but the forces that organize a galaxy cause it end up with this complex structure.

Organization is not just the location of one object in relation to another, but more specifically the movement of one object relative to another. We can say that two objects are organized when they move in relation to one another. When two elements share electrons, they become a molecule and from then on, the two elements move in relation to one another. Similarly, planets that form a solar system all move through the galaxy relative to the same sun. Organization is the relative movement of objects within some frame of reference; a galaxy and a solar system are not only levels of organization but also frames of reference. Greater organizational complexity occurs when relative motion takes place in successively greater frames of reference.

Complexity is nothing more than relative motion in multiple tiered frames of reference. Thus, a moon moving relative to a planet has less organization than a moon moving relative to a planet that is moving relative to a sun, and this has less organization than a moon moving relative to a planet that is moving relative to a sun which is moving relative to a galaxy. The planet, the solar system and the galaxy are a series of greater frames of reference in which the moon moves, and each is an additional level of organization. The level of complexity depends on which frame of reference is being considered. A galaxy is a more complex structure than a solar system. In contrast, objects that move independently of one another are not considered to be organized. If there were no galaxies or solar systems, just planets and stars moving randomly through the universe, the universe would be less complex, although probably much more confusing. Organization gives us a frame of reference in which to understand motion.

We recognize something as being an individual structure because the distribution of parts is connected by the relative motion of those parts to one another, and this relative movement distinguishes the whole of the structure from its surroundings. Organization is the relative motion of a distribution of parts, and complexity is the number of levels of relative motion, or number of frames of reference, in which a distribution of parts moves.

We tend to perceive an organism's form as being in a fixed state, but in reality, it is in a continual state of motion. Molecules are continually moving within cells, and cells are always in flux within the body. We recognize the aging that this movement causes, and we know that this internal movement is occurring, but because the external changes it causes happen so slowly, our perception of form is that it is fixed. What we may perceive as being a fixed organic form, is in fact the pattern of movement of the parts that compose a form. This movement is the process of development that creates and sustains biological form.

Form is a consequence of organized motion. Objects become organized when they move relative to one another within a frame of reference, and it is the movement of objects relative to one another that creates form. It cannot be stressed too greatly that form is fundamentally a pattern of motion. Electrons, protons, and neutrons whizzing around in confined space give an atom its form. Atoms form molecules by sharing electrons. The shared electron confines the combined atoms to moving relative to one another, and this gives the molecule its form. Molecules moving in relation to other molecules form all kinds of organic and inorganic substances. Atoms and molecules all have a form that is created by the organized movement of their substructure. We identify these structures as having individual form because the components of their structures move relative to one another as the structure moves, which distinguishes it from its surroundings.

Physical structure is organized into levels of increasingly greater organization that extends from the very small to the very large, subatomic particles form atomic particles, atomic particle form atoms, atoms form molecules, and so on up to the universe. This same kind of complex organization that occurs in physics, also occurs in biology. In biology, organization ranges from the very small organic molecules up to the biosystem, with a myriad of structures taking shape in between, and there are certain levels of structural organization that can be identified within this range of structures.

In nature organic molecules are organized into cells, cells into organisms and organisms into ecosystems. Each increase in organization represents increased complexity. Complexity increases by arranging smaller structures to form larger structures. Simple organization is a set of structures spatially organized to produce another structure. Complex organization occurs when a set of structures, each formed through simple organization, are themselves spatially organized to produce an even larger structure. The more complex a structure, the more levels of organization it takes to produce it.

CATEGORIES OF STRUCTURE

There is an idea that has been around for many decades, that an organism is an ecosystem of cells, and similarly, that a cell is an ecosystem of molecules. In this section we are going to expand on that concept to describe a biological form as an aggregate of the distributions of its constituent populations. In other words, an organism derives its form from the distribution of the cells that compose it, and a cell from the distribution of molecules that compose it. This reduces the unit of form to the individual and allows us to later describe evolution and development in terms of a population interacting with its environment.

The suggestion that an organism is a system composed of populations implies that the system's form is an aggregate of the forms of each population that composes it. The distribution of a population has a three-dimensional shape, and that shape contributes to the overall shape of the organism. So, for example, the human form is an aggregate of its various compositional cell populations and these populations each have a form that contributes to the human form. More specifically, a single muscle in one's arm gets its shape from the distribution of the population of muscle cells that compose it. The shape of this muscle contributes to the shape of the arm and to the overall human form.

There are other, more subtle, implications of this concept. First, it implies that the cell populations which form an organism are like the animal and plant populations that form an ecosystem, and therefore should develop and interact in much the same way. This suggests that ontogeny, phylogeny, and evolution follow similar rules and that corollaries can be drawn between these processes.

Second, it implies that all biological things that are composed of populations have a three-dimensional form. Such things as ecosystems and the environments that compose them, then have a three-dimensional form. It is quite difficult to envision the three-dimensional shape of an environment, yet most organisms exist by consuming other organisms. Every prey population has a distribution that contributes to the form of their predator's environment.

Moreover, the physical elements of an environment, such as water, sunlight, and minerals, also have distributions over space and time that contribute to the three-dimensional form of an environment. Both environments and populations wherever they may exist have a distribution that gives them a tangible, three-dimensional form, regardless of how difficult it may be to delineate that form.

It is possible to divide all biological structures into just three categories: individuals, populations, and systems. Evolution has organized these three structural categories into four hierarchical levels of organization: the molecular, the cellular, the eusocial and the biosystem. In this organizational structure, individuals are reproduced to form populations; populations expand, diversify, and integrate to form systems; and a system acts as an individual at a higher level of organization. This hierarchy of organization progresses form the molecule to the biosystem.

Beginning at the molecular level of organization, DNA is the reproducing molecule that initiates the production of a diverse variety of molecular units of form which are proteins. These individual molecules form populations, and diverse populations of molecules integrate to form the cell. A cell is a molecular system composed of a variety of molecular populations. Organization then proceeds at the next greater level of organization with the cell acting as the unit of form. Following the same developmental process that occurs at the molecular level of organization: cells reproduce to form populations, and cell populations diversify and integrate to form a multicellular system. The cellular level of organization organizes diverse cell populations into multicellular organisms. The multicellular organism then becomes the unit of form in the next greater eusocial level of organization. A eusocial colony is a system composed of multi-cellular organisms. Each caste within the colony is differentiated population integrated into the whole. The final level of organization is the biosystem. It is the most complex level of biological organization, and it is formed by the integration all the various kinds of organismal populations that exist. Populations of viral, cellular, multicellular, and eusocial organisms compose the various ecosystems that comprise the biosystem.

For the purpose of biological organization, the terms individual, population, and system each have a specific meaning, different from their common usage. An individual is the unit of form in biological structure. It is a reproducing or reproducible biological structure that occupies a position within its environment and can form a population. The compositional parts of an individual share a common origin in a singular developmental process that produces the entire individual. The individual develops as a whole and perishes as a whole with all its parts sharing a common demise. Biological individuals have a function determined by their form and they move both within their environment and relative to their population. An individual is a reproducible structural unit in the complex organization of biological form and throughout this book it may be referred to as an individual, a unit of form, a unit of structure, or a structural unit.

Proteins, cells, multi-cellular organisms, and eusocial colonies are all included within this definition of an individual. Tissues, organs, and ecosystems fall outside of the definition. None of these latter structures are reproducing, they do not form populations, and none occupy a position, as this term will be defined later in the discussion of space. Such things as organelles, organs, and tissues tend not to live and die as individual structures but do so as a part of a larger individual. For example, tissues are populations of cells and organs tend to be a grouping of tissues. Tissues and organs are not reproducing individuals and they do not exist on their own, but rather live and die as part of a greater organism.

Individuality is commonly determined by the motion of the individual's parts. If the parts of something move relative to one another or relative to the same thing, then they are generally considered to be compositional parts of the same entity. For example, as one moves all the cells in one's body move in unison. A person's movement causes each of his cells to move relative to the movement of all the other cells in his body.

In biological organization, all organisms – viruses, cells, multicellular organisms, and eusocial colonies – are defined as individuals, as well as the various molecules – proteins, nucleic acids, lipids, and carbohydrates – that form cells. A eusocial colony is considered an individual because each part moves relative to the whole, even though its parts are spatially disconnected and can move independently. The eusocial colony also exhibits all the characteristics of an individual. To function as a unit of form in the organization of biological structure, an individual must be reproducible as a single unit and able form a population.

There are other things often considered to be individuals because they appear distinct from their surroundings. They stand out from their surroundings, they are identifiable, and often have specific purpose. A mountain on the landscape or the nose on one's face, for example. However, these things have no identifiable boundary that can delineate them from their surroundings. Nor do they move or develop independently from the whole of which they are attached. These kinds of things, although recognized as individuals, are features of a greater whole. Perceived individuality alone does not define the individual for the purpose of biological organization.

The second of the three categories of structure is the population. An individual is defined in part by its ability to form a population. A population, as the term will be used here, is defined as any group of similar individuals that move relative to one another and occupy the same environment. Individuals are similar if they have the same form and function, and they will move relative to one another if they occupy the same environment. Similar individuals that occupy separate environments do not move relative to each other, but their populations, if in the same system, will move relative to that system. Environment delineates a population from other similar individuals by causing the individuals within it to move relative to one another and to their shared environment.

Similarity of form is necessary to delineate a population because it is quite common for different forms of individuals to be intermixed in the same area, utilizing the same resources for the same purpose. For example, wolves, bears, and mountain lions may occupy the same area and consume much the same prey for the same purpose, but they are different forms associated with different kinds of populations. Different populations and environments often overlap, and resources are often common among overlapping populations.

Form and function are inseparable aspects of structure, a variance in the ability to function always indicates some variance in form. Individuals with the same form will function in the same manner under similar conditions and react similarly to their environment. For this reason, the individuals of a population move relative to one another and relative to their environment. A population not only moves relative to its environment, but also relative to the movement of its environment.

A population is not a species, but rather a species may have several populations. A population is a group of individuals of the same species, or type, which occupies the same environment. Individuals occupy the same environment if they utilize the same set of direct resources in the same area for the same purposes. If the same variety of individuals utilize identical environments in separate regions, they do not have a shared environment. Individuals will move relative to one another only if they inhabit a shared environment, and individuals that do not move relative to one another are not members of the same population.

There can be many distinct populations of the same variety of individual within a system. For example, think of the various muscles in the human body, most of the cells have similar form and function, yet each muscle is a separate population because each muscle can move independently of the other muscles. However, the cells of each muscle move in unison. The shared environment of a population is the frame of reference in which the individuals of a population move. The environment delineates the population and causes the individuals of a population to move relative to one another and to their shared environment. This distinction will become more important when we discuss how structures develop.

The third category of structure is the system. A system is a group of diverse, interdependent populations that form a single individual that occupies a position. Cells, multicellular organisms, eusocial colonies and the biosystem all function in this manner and all are composed of interdependent populations. As an individual, a system has its origin in a single developmental process. A system is formed when individuals reproduce to form populations, and those populations diversify and integrate to form the system. The function of a system is to efficiently convert the resources of its position into environments that support its compositional populations. Because of this, the populations composing a system tend to survive to the extent that the system survives, and a system will survive to the extent that its compositional populations remain viable. A system originates, grows, and dies as an individual. Highly integrated systems, which include cells, multicellular organisms, and eusocial colonies, can reproduce. A biological system is a living, growing individual, and as an individual, a system can be a unit of form.

Biological forms are composed of both biological and physical systems. A multicellular organism, for example, is a biological system composed of individual cells that are also biological systems. A cell is composed of molecules which are physical systems rather than biological systems. While biological systems are constructed according to the rules of biology, physical systems are constructed according to the rules of physics. However, the molecules that construct cells are unique among molecules in that they can replicate themselves and form populations. The ability to replicate and to form populations are characteristics necessary to the formation of a biological system. All organisms are biological systems.

We recognize structures as having individual form because their parts move relative to one another. A population is discernible because its individuals move relative to one another, they do so because they exist in the same environment. Likewise, a system is discernible because its compositional populations move relative to one another. The form of a structure is simply the distribution of its parts that move relative to one another.

All biological structures have a three-dimensional form which is easy to recognize when the parts of a form are closely associated spatially, but more difficult to recognize when they are spatially disassociated. For example, it is quite easy to see that a single cell has form. It is also easy to recognize that a population of cells has form when they assemble into a tissue. However, it is much harder to discern the form of a population amoebas or a population of trees, and even harder still to discern the form of an ant colony or that of an ecosystem, yet these too have form.

Form is distinguished by the relative movement of its compositional parts. An individual is recognized as an individual because the units composing its structure move relative to one another. Populations are formed by similar individuals moving relative to one another and systems are formed by populations moving relative to one another. All organisms are biological systems composed of populations and all populations are composed of individuals. Individuals, populations, and systems are the three categories of structure in biological organization from which all organisms are produced.

LEVELS OF ORGANIZATION

Biological organization demonstrates a scale of increasing complexity that spans from the molecule to the entire biosystem. That scale can be divided into a series of increasingly more complex levels of organization in which each level is delineated by the formation of a reproducing system. Within each level, individuals reproduce to form populations, populations diversify, and diverse populations integrate to form a system. The system produced at the lower level of the organization then acts as the individual at the next greater level of organization.

Presently, evolution has produced four levels of biological organization. The least complex level is the molecular, which extends from the nucleic acid to the cell. The least complex unit of form is the biological molecule. There exists a myriad of biological molecules that come in four basic types: nucleic acids, proteins, lipids, and carbohydrates to name a few. Populations of various kinds of these molecules form molecular structures such as chromosomes, viruses, and organelles. When molecular populations become sufficiently integrated, they can form a cell. A cell is a system composed of interdependent populations of molecules, metaphorically a molecular ecosystem, and it represents the most complex form of molecular organization. A cell is also a transitional structure that at the molecular level of organization exists as system but at the cellular level of organization, which is the next greater level, it functions as a reproducing individual.

Just as molecules can exist as individuals, in populations, or in systems, so too can cells. The cellular level of organization extends from the cell to the multicellular organism. A multicellular organism is an interdependent set of cell populations that have diversified and integrated to form a system. In the same way that a cell is produced through the organization of molecules and may be considered a molecular ecosystem, a multicellular organism is produced through the organization of cells and is an ecosystem of cells. Like a cell, the multicellular organism is a transitional structure. It functions as both a system at the cellular level of organization and as a reproducing individual at the next greater, eusocial level of organization.

The metazoan level of organization is the next more complex level of biological organization to have evolved. Like the molecular and cellular levels of organization, the metazoan level contains individuals, populations, and systems, but rather than using molecules or cells as the building blocks of the system, the metazoan level uses multicellular organisms as its building blocks. A eusocial colony is a multi-metazoan, or multi-organismal, system. From a single reproducing individual, it produces diverse populations and integrates them into a eusocial colony. Like cells and multicellular organisms, the eusocial colony is a reproducing system. The most common representatives of a eusocial system are ant colonies and bee colonies. The various castes within the colony, because they differ in both form and function, are the differentiated populations that compose the eusocial system.

The level of greatest organization is the global biosystem, which is a singular biological system encompassing all life on earth. It is not a reproducing system, but rather a continuing system that, so far, has not encountered its complete demise. While the biosystem is not necessarily perceived as an organism, it is in fact an individual biological organism formed in the same manner as every other organism. Its development is represented in the evolutionary history of life. It does grow and expand and periodically it undergoes episodes of decline and regeneration. It functions as a single, loosely integrated, metabolic process that converts energy and biomass throughout its structure in support of itself. It is composed of all viruses, cells, multi-cellular organisms, and eusocial colonies that exit. As a system, it is the most complex level of biological organization.

These four divisions of structure delineate a hierarchy of distinct levels of increasingly complex biological organization and demonstrate a repeating pattern in which evolution proceeds. Each level of organization is composed of individuals, populations, and systems; and each level develops through a process that reproduces individuals to form populations, diversifies the populations, and integrates them to form a system.

Levels of organization are delineated by the formation of a reproducing system. Cells, multicellular organisms, and eusocial colonies are the transitional structures between levels of organization that act as both individual and system. Whether a transitional structure is deemed an individual or system depends on the level of organization in which it is being considered. A cell for example is molecular system, but it can also act as an individual cell within a population of cells or within a multicellular organism. The transition from one level to another occurs when a system evolves to such an extent that it can reproduce as an individual. Once this occurs, it can then function as an individual in the formation of a greater system. Cells, multicellular organisms, and eusocial colonies are reproducing biological systems that can act as individuals in the formation of greater level systems.

Within this hierarchy of structural organization there exists a variety of different kinds of intermediate structures, as well as intermediate organisms. A virus is an intermediate type of organism that falls between the biological molecule and the cell. Viruses are composed of populations of molecules, they reproduce, they form viral populations, and they may in some sense be considered molecular systems. Mitochondria, which appear to have originated as some type of virus, along with other organelles are also a form of intermediate molecular organization composed of populations of molecules. These are produced within the molecular level of organization, as are prokaryotic and eukaryotic cells which are reproducing molecular systems. Unlike eukaryotic cells which can integrate sufficiently to form greater level systems, viruses and prokaryotic cells appear unable to do so.

Similar to organelles, organs are also intermediate structures but within the cellular level of organization between cells and multicellular organisms, and the tissues which compose them are cell populations.

At the metazoan level of organization, the various castes within a eusocial colony are the differentiated populations that form the eusocial system. The queen, which is often a population of one, acts as the reproducing organ for the system.

An ecosystem is an intermediate structure formed within the biosystem. The biosystem is an aggregate of ecosystems, and an ecosystem is composed of groups of various populations that interact trophically. While an ecosystem is certainly a system in the general sense, and it is a trophic structure as are all biological systems, an ecosystem is not considered a system within the hierarchy of structural organization by the definition used here. Unlike cells, multicellular organisms, and eusocial colonies, an ecosystem does not originate and perish as an individual. Ecosystems are neither individuals nor reproducing systems and are not transitional structures between levels of organization. Rather they are a feature in the structural organization of the biosystem, much like organelles are in cells and organs are in organisms.

Although, ecosystems are easily recognizable, it is nearly impossible to delineate an ecosystem as a single entity. This is because not only do the boundaries of ecosystems often abut and bleed into one another, but populations often migrate through different ecosystems, making a single population an integral part of more than one ecosystem. Moreover, the functioning of an ecosystem often tends to affect the functioning of adjacent ecosystems, as well as be affected by them. An ecosystem is more a conglomerate of invasive species, originating at different times and places, which have come together over time in the same geographic area to form a trophic system; it is an area where certain kinds of forms are more concentrated within the greater biosystem.

Ecosystems are produced from interacting populations of molecules, cells, multicellular organisms, and eusocial colonies. Like organelles and organs, ecosystems are intermediate structures in the organization of the biosystem that do not develop or exist as individual units and do not reproduce or form populations. However, unlike organs and organelles which perform specific functions in support of their system, ecosystems, although integrated into the biosystem, do not perform any specific function for it.

Cells, multicellular organisms, and eusocial colonies are highly integrated, reproducing systems, in which the populations that compose them interact symbiotically. The biosystem is a much more loosely organized, non-reproducing system in which its populations interact primarily through predation. Predation and symbiosis can be viewed as the two extremes on a scale of interaction between the populations that form a system.

Most organisms are systems and quite a few are complex systems. Viruses are almost systems. They appear to have all the properties of a system, but they cannot function as a system until they find a host. Life is full of organisms that seem to be halted in mid-transition from one form to another. Cells are simple systems because they are not composed of other biological systems but of molecules. Although molecules may be considered physical systems, they are not biological systems. Cells are composed of populations of molecules, and in our scheme of biological structure, a molecule is the simplest biological individual. Multi-cellular organisms, eusocial colonies, and the biosystem are complex systems because they are composed of populations in which the individuals are systems.

Biological organization ranges from the molecule to the biosystem and it exhibits a hierarchy of structural organization. The levels of organization within this hierarchy are delineated by the formation of cells, multicellular organisms, and eusocial colonies. These structures are both system and reproducing individual and act as transitional structures between levels organization. The levels of organization represent a hierarchy of greater levels of complex organization, and the evolution of this hierarchy suggests that the biosystem evolves toward greater complexity.

Within each level of organization, development follows the same process which begins with individuals reproducing to form populations followed by the diversification and integration of those populations to form a system. This similarity in the sequence of organization among the levels of organization suggests that the process of development may also be consistent across all levels of organization. This indicates that embryonic development and the evolutionary diversification of species are the same processes acting at different levels of organization.

UNITS OF FORM

At each level of organization, biological systems are constructed using the three categories of structure: individuals, populations, and systems. Systems are composed of populations, and populations are composed of individuals. The individual is the primary unit of structure, a population is an intermediate unit of structure in the formation of a system, and a system can be a unit of structure if it acts as a reproducing individual in the formation of a greater population or system.

Molecules, cells, multicellular organisms, and eusocial colonies all act as individual units of structure in the formation of higher-level systems. Molecules are units of structure in cells; cells are units of structure in multi-cellular organisms; multi-cellular organisms are units of structure in eusocial colonies. Viruses, single-celled organisms, multicellular organisms, and eusocial colonies are units of structure in the composition of the biosystem.

Except for the molecule, individual units of biological structure are all biological systems. Every organism is some form of biological system and biological systems are constructed by producing individuals, organizing them into populations, and organizing the populations into a system. The system produced is then used as the unit of structure at the next greater level of organization. Greater levels of organization are produced with lower-level systems acting as their individual units of structure.

The individual is the basic unit of structural organization and biological form is the phenotypic expression of structural organization, so the unit of structure is also the unit of biological form. Structural unit, unit of structure, and unit of form are interchangeable terms with essentially the same meaning.

Identifying the units of structure completes the system of structural classification that has been described throughout the preceding sections. This system is presented in the following chart, and it is a system that recognizes complexity rather than inherited relationships.

Level of organization	Unit of Form	System Type	Examples
Molecular	Molecules: nucleic acids proteins, carbohydrates, lipids	Cell	Prokaryotes & eukaryotes
Cellular	Eukaryotic cells	Multicellular organism	Plants & animals
Metazoan	Metazoan Organisms	Eusocial colony	Eusocial colonies
Biosystem	All organisms: viruses, cells, plants, animals and eusocial colonies	Biosystem	The biosystem

Reducing the myriads of biological forms down to three categories of structure within a hierarchy of levels of organization and identifying the individual as the basic unit of structure simplifies the complex process of development. One may assume that form develops in the same manner within each level of organization and that development within each level organizes individuals into populations and populations into systems. Unravelling the process of development can then be reduced to answering four questions, each related to the four elements of form – variety, number, location, and timing. What determines variation in individual form and causes a population to diversify? What determines the quantity of individuals within a population? What determines the location of an individual and the distribution of a population? What triggers the timing of developmental events? To produce an organism, a developmental process must produce a variety of individuals, each in the correct quantity, and distribute them in the correct location at the appropriate time, often at several levels of organization simultaneously. These questions are interrelated, and their answers can be found in the dynamics of a population and the functioning of individuals and systems.

VARIETY

The phenotype of an organism has a form that is produced by the three-dimensional distribution of its parts. For example, the form of a multicellular organism is the distribution of its cells, its cells are organized into populations, and its form is an aggregate of the distributions of its cellular populations.

An organism's form may be described using the same method that architects and engineers use to describe buildings and machines, and that is to use a blueprint. A blueprint specifies three parameters, the variety of parts, the number of each kind of part, and where the parts must be placed in relation to each other. Mechanical engineers use this method when describing machines. An expanded diagram of the machine shows the variety, number, and location of the parts that compose the machine. Such diagrams are often shown in the owner's manual of common household appliances.

The structure of an organism may be defined in this same method by using only the three parameters variety, number, and location. For example, a multicellular organism can be described by identifying the variety of cells that compose the organism, the number of each type of cell, and the location of each cell in relation to the other cells in the organism. While this method may be impractical for identifying organisms, it is none the less very accurate.

The cells within an organism are grouped into populations and each of the organism's tissues is a population of the same type of cell. An organism is a system and while it requires three parameters – variety, number, and location – to describe the form of a system, it requires only two parameters – number and location – to describe the form of a population. This is because a population is composed of only a single variety of individual. A population has a distribution that is determined by the number of individuals in the population and their locations relative to one another. This distribution gives the population a three-dimensional form, and that form contributes to the form of the system. A system's form is then a composite of the distributions of its constituent populations.

The distribution of a population is determined by the number and location of the individuals in the population and a change in either parameter can change its distribution. Because the distribution of a population contributes to the form of its system, a change in a population's distribution will cause a variation in the form of the system. A variation in the form of a system can then arise from a change in just these two variables of form: number and location.

The parts of a system are individuals, which are distributed as populations, and altering the distribution of the parts a system, even without changing the variety of those parts, can alter the form of the system. Furthermore, if there is a sufficient change in the distribution of a sufficient number of populations within a system, the variation can alter the functioning of the system.

Consider the structure of two multi-cellular organisms: a man and a mouse. Both are composed of the same approximately three hundred or so varieties of cells, each variety has approximately the same form and function in both organisms. The two forms differ, not because they are composed of different varieties of cells, but because the cell varieties have different distributions in each organism. The man and the mouse differ by only two parameters of form. The number of each kind of cell and the location of the cells, these differences likely arise from a difference in the timing of developmental events, but the variety of cells is the same in both organisms. Altering the distribution of one or more cell populations within and organism modifies the shape of the tissue and in turn the form of the organism.

This becomes even more evident when we consider the evolution of closely related species such as horses and donkeys or deer and elk. Deer and elk are distinctly different forms, and therefore functionally different. However, their differences in form arise from the different distributions of their compositional cell populations. The same is true of horses and donkeys, which are so closely related that they can be interbred. This principle was depicted graphically by D'Arcy Thompson in his book "On Growth and Form." His illustrations show how form evolves by distorting various aspects of form. Such distortions arise as evolution modifies the spatial distribution in the compositional cell populations of a descendant from that of its ancestral form. The evolution of elk and deer from their common ancestor requires only a change in the distribution of cells to make the transition from ancestral form to descendant form, not necessarily a change in the genetic template from one to the other. This becomes even more evident when we consider the various castes of ant within an ant colony.

An ant colony is composed of different castes of ant, typically queens, drones, and workers. Each caste is a differentiated population within the eusocial system; each has a particular form and function; and each form is distinct from the others. Although, the queen is diploid, the other castes are haploid and genetically identical. The phenotypic difference between castes arises from the kind of enzyme fed to the pupae, different enzymes produce different forms of ant. The difference in form of the ant castes is a result of differences in the organization of cells, rather than a result of a genetic differences between the castes.

Altering the distribution of the substructure to change the form of the superstructure is evident even in the molecular structure of amino acids. Although amino acids are composed of individual nucleotides rather than populations, variation in amino acid structure is a result of differences in the spatial organization of the nucleotides. For example, using only two types of nucleotide bases, uracil (U) and cytosine (C), four functionally different types of amino acids – serine (UCU), phenylalanine (UUC), leucine (CUU) and proline (CCU) – can be formed by changing the number and location of the nucleotides within the string. If only the order is changed, but not the number of each type of nucleotide, then three types of amino acids – serine (UCU), phenylalanine (UUC) and leucine (CUU) – can still be produced using just one cytosine and two uracil nucleotides. By simply changing the spatial organization of the nucleotides, functionally different varieties of amino acids are produced.

The division of biological form into three categories of structure, not only allows us to identify the distinct levels of organization in which organisms are produced, but also allows us to recognize that the structure of an organism is a result of the spatial distribution of the individuals in the populations composing its substructure. A variation in a form is then produced by a difference in the distribution of one or more of the populations that compose the form. This principle applies to each level of organization, cells are composed of populations of molecules, and multicellular organisms are composed of populations of cells. The form of an organism is the spatial distribution of its compositional populations, and a variation in organismal form is a result of a change in the distribution of one or more of its compositional populations.

It is important to note that variations caused by changes in the spatial distribution of a population do not manifest in the form of the individuals within the population distributed, but rather in the allometry of the system in which the population resides. The allometry of a system varies when the spatial organization of its substructure is altered, even though the form of the individuals which compose its substructure may remain the same.

From the molecular level of organization on up to the biosystem, variety is produced by altering the spatial organization of one or more of the populations composing the substructure of the form. The allometric differences that distinguish the various related species are generally the result of the way in which cells are spatially configured to produce a particular form, and not a result of differences in the variety of cells that compose the form. A cell's variety is determined by the spatial organization of its molecular substructure. Within most taxonomic families the evolution of a new form requires only that development reorganize the same kinds of cells.

Because an organism is a system composed of populations, a variation can be defined as a difference in the distribution of a population that alters the form of its system, but which is not sufficient to change the functioning of the system. Variations accumulate over generations to produce adaptations. An adaptation can be defined as a change in the distribution of a population that causes a change in the form of its system which is sufficient to cause a change in the functioning of the system. Adaptations produce varieties, while variations do not, and an organism is recognized as a different variety by the adaptations it possesses.

Because variety is a consequence of changes in the spatial distribution of the compositional populations that form an organism, the four parameters of form – variety, number, location, and timing – are in effect reduced to three. Variety is determined by distribution which is the number and the location of individuals. Every population has a distribution and whatever determines the distribution of a population determines the form of the population and indirectly the form, or variety, of the system produced by that population. By knowing what determines a population's distribution, we may then understand what determines the form of a system.

3. SPACE: Environment and Position

If the form of a system – its phenotype – is determined by the distribution of the individuals in the various populations that compose it, then what is it that determines the distribution of those individuals? The seemingly obvious answer is that the distribution of a population is determined by its environment. However, the hierarchy of structural organization described previously complicates this correlation so that a natural environment, as the term is commonly understood, determines form in only a very general and indirect manner. To clearly make the correlation between environment and distribution, environment must be defined in such a way that it corresponds to a specific population within the hierarchy of structural organization This necessitates establishing a more detailed concept of environment, and yet an environment must remain in its most basic sense a set of resources which a population converts into more population.

ENVIRONMENT

In the parlance of biology there may be no term more loosely defined than environment. Anything that may affect a population has been referred to as an environment. The resources that support a population, the resources from which those resources are derived, climate, terrain, the geophysical earth from which comes climate and terrain, predators, diseases, other populations, and the ecosystem, may all be considered as one's environment. The interconnectedness of life, along with the earth's physical structure and properties, allows nearly everything on or off the planet to be included in a definition of environment if one so chooses. It is difficult to relate a specific form to a general environment, especially if environment is so broadly defined.

Such a comprehensive concept of environment cannot be associated in any meaningful way with a population when populations exist in successively greater levels of organization. The environment in which an organism resides is not the same as the environment in which a molecule, within a cell, within that organism resides. Not only do different populations within the hierarchy of structure have different environments, but different populations within the same system have different environments.

However, it is possible to correlate the distribution of a population to an environment if that environment is defined in a way that is specific to the population. To do this, environment must be subdivided in such a manner that it can be correlated to a specific population within a system, and at a specific level of organization within the hierarchy of biological structure.

For that purpose, an environment will be defined as only those direct resources that a population has access to. This definition requires that the term "direct resource" also be defined. Resources can be divided into two categories: direct and indirect. A direct resource is one that requires no further modification to be utilized by a population for its subsistence and reproduction. Indirect resources are those resources from which direct resources are derived. Indirect resources must undergo some further modification before a population can use them.

As an example of a resource, consider a sheep. For a carnivore, such as a wolf, a sheep is a direct resource. It is also means of converting grass and other indigestible vegetation into something the wolf can digest. Vegetation can support wolves if it is first converted into sheep. The vegetation consumed by the sheep is an indirect resource to the wolf but a direct resource for the sheep. The volume of wolves is directly dependent on the volume of sheep, and the volume of sheep is directly dependent on the volume of vegetation, but the volume of wolves is only indirectly dependent on the volume of vegetation. For the wolf, sheep are a direct resource and the vegetation that supports the sheep is an indirect resource, not a direct resource.

Suppose the carnivore is a person. The sheep then becomes an indirect resource because it must be converted into something more digestible for human consumption. A person, unlike a wolf for example, is not physically equipped with the jaw strength and teeth to directly consume a sheep, the sheep must be modified into something more palatable, such as a lambchop. The person does this using his cutting tools.

Now suppose there is a population of people that subsists on only hamburgers. The volume of hamburgers available to the population would directly affect the volume of people in that population because their subsistence and reproduction is directly dependent on the consumption of hamburgers. The number of people is only indirectly dependent on the number of cattle from which the hamburgers are produced. Cattle are an indirect resource to people because the people are dependent on cattle only to the extent that the cattle can be converted into hamburgers. Hamburgers are, in this instance, direct resources and cattle are indirect resources.

If this is taken a step further and the person is not considered as an individual but rather as a multi-cellular system composed of various populations of cells, the hamburger then becomes an indirect resource for those cell populations that compose the person. The hamburger must be broken down into usable fats, proteins, and sugars that can be directly ingested into a cell. The person, as a multicellular system, catabolizes the hamburger into molecules and distributes those molecules as direct resources to the various cell populations that compose the person. A hamburger is a direct resource for the individual that eats the hamburger, but it is not a direct resource for the cells which compose that individual. The hamburger must be converted through catabolism into molecules that the organism's cells can ingest, those molecules then become the direct resources of the cell's environment.

By defining an environment as only those resources that a population can directly utilize and including only those direct resources that the population can access, an environment then becomes the total volume of energy and material that supports a population's subsistence, growth and reproduction, and nothing more. An environment is only that set of direct resources that the individuals of a population convert through their metabolism into more population. Using this definition of environment, we are then able to associate a specific population within the hierarchical organization of biological structure to a specific environment, and this allows the volume of resources contained within an environment to be correlated to the quantity and distribution of population supported by that environment.

The function of a biological system is to convert resources and distribute them as environments to their compositional populations. This is what the trophic levels of an ecosystem do, it is what the metabolism of an organism does, and it is the purpose of all cell functions.

Indirect resources, inaccessible resources, and such things as predators and disease are not included in one's environment because, while these may affect a population in particular ways, a population is not directly produced from these things.

Every individual has a form that allows it certain abilities which enable it to access and convert resources. Form evolves by accumulating adaptations, and an individual's form represents the accumulated adaptations that it has retained throughout its evolutionary history. These adaptations determine its ability to convert resources, and it cannot exist without doing so. An individual's form is directly related to the direct resources it converts and it is configured in a way that it can acquire, utilize, and consume the direct resources of its environment. Although, the same resources may be used by different kinds of organisms and different populations, each population utilizes a specific set of direct resources that are its environment.

At each level of organization, distinct populations have distinct environments and each population within the hierarchy of system can utilize only its direct resources. An environment is a finite set of direct resources that includes both organic and inorganic material, and it is the source of the energy and material that produces and sustains a population. An environment has a distribution that fluctuates in volume, space, and time and that distribution confines the movement of a population to within its boundaries, causing the distribution of a population to fluctuate along with the distribution of its environment.

By defining an environment as a population's direct resources and including only those resources that a population can access, environment is limited to the set of resources that a population directly utilizes to produce and sustain itself. A volume of environment can then be correlated to a volume of population, and a volume of population has a distribution that contributes to the form of the system in which the population resides.

POSITION

Distinguishing direct resources from indirect resources, not only delineates the distribution of an environment, but also identifies that set of resources which corresponds to a particular population. A population further subdivides its environment into positions. While an environment is the set of direct resources available to a population, a position is the set of direct resources within an environment that is available to a single individual. Position is an individual's access to the resources of its environment, it delineates the total volume of direct resources available to a single individual, and it is the subset of resources within an environment that is accessible to a single individual. No two individuals can occupy the same space, nor can they consume the same resource, therefore a position is exclusive to an individual. A position is that portion of an environment that corresponds to the individual.

An environment is the volume of direct resources available to a population and it confines a population to moving within its distribution. Position is a subset of those resources within an environment that is available to an individual, and like an environment, a position has a distribution that confines the individual to moving within that distribution. Because position is the division of an environment among the individuals of a population, the number of positions within an environment is always equivalent to the number of individuals in the population. Both environments and populations fluctuate continually, causing the positions within an environment to fluctuate in both size and number. A position provides the set of resources necessary to support an individual, so throughout an environment, positions are similar in the variety of resources available but may differ in the volumes of those resources. Positions tend to be relatively similar and stable in mature populations, while in developing populations they tend to be more varied.

A position may endow an individual priority in access to resources over other individuals of its population. That priority may be due to location, timing, or social dominance.

Consider for example a row of trees that receives a limited supply of water from an inlet at one end of the row. Because the soil becomes less saturated the farther a tree is from the water source, the amount of water that a particular tree receives depends on how far it is from the water source. Thus, the location of a tree in relation to its environment will affect its development. In addition, the amount of water that each tree closer to the water source uses, will affect the amount of water that remains available to those trees farther away. Were there no trees between the water source and the farthest tree, presumably more water would reach the farthest tree. Each tree's access to water is determined not only by how far it is from the water source but also by how many trees have access to the water before it. In this example not only does the individual's location in relation to its environment affect its development, but so also does the individual's location in relation to other members of the population.

Now suppose that the row of trees has matured and a tree in the middle of the row is replaced with a sapling. The trees on either side of the sapling have mature roots and branches that encroach on the position of the sapling, so that at all phases of its growth cycle the sapling is allowed less water and sunlight than the tree that preceded it. This demonstrates how timing can affect an individual's access to resources at a position. Position is an individual's access to the resources of its environment and that access depends on the distribution of the environment as well as the use of resources by the population, and both can change over time.

As these examples suggest, all positions are not the same and some positions may endow their occupant with an advantage in accessing resources. This advantage comes from the priority that one position provides over other positions. Priority can be spatial, temporal, or social. The first scenario in the example above describes spatial priority - those trees closest to the water source get more water than those further away. The second scenario in the example illustrates temporal priority – those individuals arriving at an environment earlier will generally have more access to resources than those arriving later. Social priority tends to occur only in those animal populations that exhibit a social order. The physical dominance of one individual over another often bestows greater access to resources on those individuals exhibiting that trait. Spatial, temporal, and social priorities, for whatever reason they occur, create variances among the positions within a population.

A position is the unit of environment that corresponds to an individual and it is the total of resources that are available the individual. If the individual is a system, then position is the total of resources available to the system.

Systems are individuals, and as individuals they occupy a position within their population and environment. For example, a multicellular organism is an individual in its population of organisms, but it is also a multicellular system composed of various cell populations. The cell populations that compose the organism exist in environments that are derived from the resources available to the organism at its position. The direct resources of the organism's position are the total of indirect resources that are available to support the cell populations composing the organism.

In the hierarchy of biological organization, cells, multicellular organisms, eusocial colonies and the biosystem are individuals that are also systems. As individuals, these systems occupy a position within their environment, and that position is the entirety of resources available to the system.

The biosystem represents the macro end of the scale of structural organization and, unlike cells, multicellular organisms, and eusocial colonies, it is not an individual within a larger population, rather it is an individual that occupies its entire environment as its position. It is a singular system that resides within the geosphere, and the geosphere acts as both position and environment for the biosystem. At the micro end of the scale are the biological molecules that form cells. These individuals have positions within their environments, and they form populations, but they are molecular not biological systems.

The function of a biological system is to convert the resources of its position into environments that can support its compositional populations. Those populations then subdivide their environments into positions, and the process of converting positional resources into environments is repeated at the next sublevel of structure. This is repeated down through each level of organization within the biosystem to the molecular level, and it is essentially the pathway though which a system catabolizes resources. At the molecular level, anabolism begins assembling resources into the individuals, populations and systems that form all biological structures.

Consider the resource flow in a eusocial colony. A eusocial colony is an individual within its population of eusocial colonies that resides in an environment within the biosystem. Each colony occupies a position within that environment, and that position is the total of direct resources that the colony can access within its environment. The colony converts the resources of its position into environments that sustain the various caste populations composing the colony. Each caste within the colony represents a differentiated population. Each individual occupies a position within one of those populations, and its position is the total of resources available to support that individual. The individual is a multicellular system that converts the resources of it position into environments that support its compositional cell populations. Each of its cell populations exits within an environment that is subdivided into positions, and each position is the total of resources available to a single cell. Each cell is a molecular system that converts the resources of its position into molecular environments that produce and sustain the cell's molecular populations.

This process of converting positional resources into environments and dividing environments into positions, is repeated down through the levels of organization, and it distills the resources of the biosystem down into the distinct molecular environments that exist within each cell. The division of environments into positions determines the volume of resources available to the individual at each position, and the conversion of positional resources by the individual into sub-level environments determines the variety and volume of resources available to its compositional populations. This is catabolism, which creates at each level of organization environments specific to each population and positions unique to each individual.

While catabolism converts positional resources into environments, anabolism converts positional resources into population.

Beginning at the molecular level of organization, position plays a primary role in determining differential genetic transcription. The chromosomes are a molecular population of genes within the structure of a cell. Genes are similar individuals, with the same structure, the same function, and they exist in the same nucleoplasmic environment. Each gene occupies a stationary position within its population, and while genes do not move relative one another, their nucleoplasmic environment does circulate. When and how often a gene may be transcribed is determined by the resources available to it and by the regulatory genes in its population that may affect it. The circulation of the environment can cause positional resources to vary over time, and regulatory genes may inhibit or promote transcription depending on their location and the content of the environment. This implies that a gene's ability to be transcribed is determined by its position, both relative to its environment and relative to its population. However, the concept of position goes somewhat further to suggest that all genes are, to some extent, regulatory genes regardless of how undetectable their regulatory effect may be. In this view, the genetic population may act as either catalyst or inhibitor on the action of individual gene depending on each gene's location.

In a finite environment, which all environments are, the use of resources by one individual precludes other individuals from utilizing that same resource. Genetic transcription requires resources, so the act of replicating done by any one gene impacts the availability of resources to all other genes. Thus, the functioning of one gene by utilizing its environmental resources to replicate itself may diminish the volume of resources available to another gene, limiting that gene's ability to replicate. Which, when and how often a gene is transcribed depends on its environment and the genes around it, this is the effect of position on the gene.

The genes initiate anabolism and begin the formation of the molecular populations that make up a cell. Cells are composed of carbohydrate, lipid, protein, and nucleic acid molecules, all of which have many varieties and form the populations that compose a cell. The reproductive populations in the cell are composed of nucleic acids, and these produce proteins which are the functional populations that carry out the cell's metabolic activity. Lipids and carbohydrates form the cell's structure and store the cell's resources. All the various types of these molecules are organized into populations with each individual molecule occupying a position within its population. Each molecule's ability to function is determined by the resources available to it at its position.

Once molecular anabolism has produced a cell, then cellular anabolism can construct a multicellular organism. Like the molecules in a cell, there are different varieties cell populations that compose a multicellular organism. Each cell has a position within its population and its ability to carry out its metabolic function is determined by the resources available to it at its position.

Position is determined by the location of the individual relative to its population. A position is that portion of an environment's resources that are available to an individual. It is determined by the availability of resources in the environment and by the usage of resources by all other members of the population. Each position within a population affects, and is affected by, all other positions. Position is the individual's access to it environment, and it delineates the space in which an individual moves. Because an environment is a finite set of resources that restricts a population to moving within its boundaries, an individual moving within its position will always be moving relative to its environment and to its population.

THE CONVERSION OF MATTER

If one considers the fields of study that are physics, chemistry, and biology, it should not go unnoticed that chemistry is a subfield of physics or that biology is a subfield of chemistry. Physics is the study of matter, energy, and the structure of atoms, while chemistry deals primarily with elements and the structure of molecules. Biology concerns itself with only that small subset of molecules, within the vast universe of molecular structures, which are amino acids. A few of these amino acids have the unique ability among molecules to bond into chains in which the pattern of bonding can be replicated. Out of these few molecules arises all biological organization.

A chemical reaction, at its very essence, is a process of converting one form of matter into another by changing its structure, this is also the essence of biology. Elements and molecules react with one another in specific ways depending on their atomic structure, and the reaction reorganizes matter by breaking chemical bonds and forming new ones. The product of a chemical reaction is always a substance, or substances, which are structurally different from the reactants entering the reaction. Likewise, biological development also converts matter from one form into another by breaking and reforming chemical bonds, however, the structure of the matter produced often replicates the initiating reactant.

Biological development can be viewed as an extended chemical process. While a simple chemical reaction can occur singularly between two elements, replicating biological reactions always occur in a series. Much like industrial manufacturing where a sequence of chemical processes is set up to produce a specific result, biological molecules undergo a sequence of reactions to replicate themselves. The amino acid sequence of a gene is replicated by first producing its RNA complement. The RNA molecule then becomes the template for replicating the original amino acid sequence of the gene. This newly replicated sequence of amino acids along with some added molecules form a polypeptide. This initial process of replication is not a single chemical reaction but rather an ordered series of chemical reactions, and all biological replication is achieved by extending this series chemical reactions.

Proteins are formed by bonding together various polypeptides. Protein synthesis is the basis of cellular reproduction and cellular reproduction is fundamental to an organism's reproduction. This causal link between a chemical reaction and biological reproduction allows biological reproduction to be thought of as a greatly extended series of chemical reactions. And, of course, we have a name for that extended series of reactions, it is "anabolism." Anabolism is the part of metabolism that builds biological structure, and it is essentially an extended and complex series of chemical reactions that builds proteins, cells, and organisms.

Anabolism has its counterpart in catabolism. Catabolism is the deconstructive part of metabolism that breaks down environmental resources as they are consumed by the organism and converts those resources into the raw materials that anabolism needs to build biological structure. Anabolism is sustained by catabolism and catabolism is sustained by resources contained in the organism's environment. Metabolism is the entirety of these two processes which an organism employs to convert its environmental resources into biological structure. Those environmental resources are mostly other organisms, but they may also be minerals, sunlight, and other matter. In fact, the sole function of an organism is to convert its environment into population by consuming and utilizing its environmental resources to sustain and reproduce itself. In this respect, a population and its environment may be considered to interact in much the same manner as reactants in a chemical reaction. Metabolism is the reaction between a population and its environment, and the product of this reaction is more population.

In chemistry a reaction technically refers to the breaking and re-forming of a chemical bond. But in a more general sense, it is essentially the disassembly of one kind of matter and the reassembly of it into another kind of matter. In biology, the process of disassembling and reassembling matter is metabolism. So, a chemical reaction and biological metabolism perform the same function – the conversion of matter – in their respective domains. However, this is not a correlation between two different processes but rather an extension of the chemical process into the more complex process of biological organization. At its most fundamental level, metabolism is an extended series of chemical reactions, and each organism's metabolism entails a vast multitude of these series of reactions.

The most basic process of metabolism is genetic replication, which is entirely a chemical reaction. The most extended process of metabolism, which arises from genetic replication, is the reproduction of an organism. Progressing from one to the other is matter of compounding the number of chemical reactions in the series. We tend to think of the chemical reactions that enable genetic transcription as quite different from the reproduction of an organism, but they are in fact the same process. The reproduction of an organism is just a much more extended and complex series than genetic replication, yet both are processes that convert environmental matter into biological form, and the former process is wholly dependent on the latter.

Economics is not normally considered to be a subfield of biology, yet in very much the same sense as biology is a subfield of chemistry, economics is a subfield of biology. Similar to the way in which the entirety of biology extends out of a single class of molecule, economics is practiced by only one biological organism out of the billions of species that have existed. But this is not what connects the fields. Chemistry, biology, and economics are fields that each study a specific type of conversion process in which one form of matter is converted into another. In chemistry, elements and molecules react with one another to form other elements and molecules. In biology, organisms metabolize their environments to sustain and reproduce themselves thereby converting environment into population. Economics is the practice of using means external to the organism to convert resources into a form that can be metabolized by the organism.

Except for a few minor exceptions, all organisms, other than humans, must rely solely on their physical form to directly acquire, consume, and metabolize their environmental resources. This is the single function for which the design of an organism's form evolves. In contrast, humans may employ a wide variety of tools and methods to convert resources into a form which they can metabolize. Economic systems extend metabolism and exist for the same purpose of converting resources into population. For this reason, economic activity may be thought of as extra-biological metabolism. The concept of conversion, in its most basic meaning, refers to the reorganization of matter, and Chemistry, biology, and economics are the study of different categories of conversion processes.

The function of an organism is to convert its environment into more of itself, either through subsistence or reproduction. Metabolism is generally thought of as a developmental process rather than as a conversion process, but as a conversion process, metabolism is a continual and repetitive pattern of movement that organizes units of structure at the different levels of organization into more individuals. If biology is to be assigned a purpose, that purpose is to convert resources into population. Every structure within the hierarchy of biological organization is designed for the purpose of converting resources or facilitating the conversion of resources.

Moreover, converting resources requires movement, both by the individual converting the resource and by the resource being converted. An individual acts to acquire and metabolize resources and those resources are disassembled and reassembled into units of structure. All biological movement, whether it be to acquire or to utilize a resource, is for the purpose of converting resources. Biological motion may in this sense be considered as synonymous with the conversion of resources.

4. MOTION: The Dynamics of Form

An organism cannot move beyond its environment for any longer than it can be sustained by the resources it may carry with it, and few organisms carry their environmental resources with them. A person for example cannot stay underwater for any longer than he can hold his breath, unless he builds some kind of artificial environment. An environment therefore confines a population to moving within its boundaries.

The distribution of a population conforms to the distribution of its environment and a population's distribution has a shape that contributes to the form of the system in which the population resides. This means that the shape, or allometry, of a system is derived from the shape of its populations' environments. In this section we will look at the ways in which environment and position direct the movement of individuals into an organized distribution of population.

CIRCULATION AND FLOW

All units of form can move and there are two basic kinds of motion by which individuals and populations move, circulation and flow. Circulation is the movement of individuals from one location to another within their environment. Circulation increases when more individuals move per period of time and decreases when fewer move per period of time. Flow is the movement of individuals into and out of a population through reproduction and mortality. Flow increases when more individuals replace fewer individuals and decreases when fewer individuals replace more individuals. Circulation is a change in the distribution of the same individuals over time and flow is the exchange of individuals over time. Both types of movement may cause a change in the spatial organization of the population. Movement through circulation and flow allows the distribution of a population to fluctuate along with the fluctuations in its environment.

In a natural ecosystem, animal populations tend to circulate and flow while plant populations tend only to flow. A migrating herd or a school of fish swimming, are examples of circulation. Animals circulate within their environment toward the resources that support their existence. Plants, which tend to be stationary organisms, move primarily through the process of flow. As individual plants reproduce and die, the geographical distribution of their population may change. Through flow a population of trees, for example, may move across the surface of the earth following a changing climate. As new trees are reproduced at the leading edge of the forest and old trees die at the trailing edge, the form of the forest changes shape and moves. As is evident in the movement of a forest, flow may occur slowly over a very long time.

At the multicellular level of organization, cells both circulate and flow. Blood circulating through veins and arteries is an example of circulation in a cell population. In this example circulation is caused by the mechanical action of the heart. Flow occurs when cells reproduce an die. As the somatic cells in our body reproduce and die and over time, a flow is created that we recognize as growth and aging. At the sub-cellular level, the molecular building blocks of the cell also both circulate and flow. At this level, the circulation of these building blocks is accomplished through chemical affinities, molecular gradients, and the expenditure of energy in the form of ATP.

Circulation occurs when a population is mobile and if its environment allows, or forces, movement. It also tends to occur in defined patterns such as a migratory path, or within the confines of blood vessels, or along chemical gradients as suggested in these examples. Circulation occurs within a very short time relative to flow.

A blood cell may circulate many times through the body before flow replaces it with a new cell.

All populations are in continual motion, moving toward their environmental resources. Movement enables a population to access its environmental resources, and organisms cannot survive unless they continually move toward their resources. Just as a population moves toward its environmental resources, individuals always move toward the resources of their position. Stationary organisms, such as most plants, which move only through flow, always move to positions where environmental resources can circulate toward them. The rule that a population always moves toward its resources means that biological movement is directed by its environment.

Resources, like the populations that utilize them, also continually move, which means that the circulation and flow of a population will mimic the circulation and flow of its environment, causing the distribution of a population to conform to the distribution of its environment, and to the fluctuations in that environment. The distribution of a population has a shape, and that shape reflects the distribution of its environment.

THE BIOLOGICAL REACTION

Recognizing that each population within the hierarchy of biological structure inhabits an environment of direct resources specific to that population, allows a volume of environment to be correlated directly to a volume of population in the same manner that specific volumes of chemicals in a chemical reaction correlate to a specific volume of product from that reaction. However, the nature of biological reproduction does not allow the exactitude in biological reactions that can be calculated in chemical reactions. Regardless, the same principle exists, and it is the primary limit on development.

Biology is fundamentally the process of converting environment into population, this is accomplished through chemical reactions, replication, and reproduction. Biological replication and reproduction have essentially the same meaning, and they refer to an extended series of chemical reactions that organize catabolized resources into a biological form. Like chemical reactions, they are conversion processes and subject to the same physical laws as chemical reactions. Hereafter, replication and reproduction will be referred to as biological reactions, where the reaction taking place is between a population and its environment.

When two chemicals are combined in a chemical reaction their units of structure change their patterns of motion causing the structure of the reactants to disassociate and recombine in a different configuration. In chemistry the units of structure are electrons, protons, and neutrons. The form of an element or molecule is the pattern of motion of its subatomic units of structure. When molecules react, the pattern of motion of their electrons change, this causes the structure of the reacting molecules to disassociate and recombine to form different molecules. The breaking and forming of chemical bonds is essentially a change in the pattern of motion of the electrons. Metabolism does much the same by breaking down an environment and reorganizing it into population. Just as the term "chemical reaction" is a generalization for the altering of electron orbits at the atomic scale that occurs in a chemical reaction, describing a biological reaction as a population converting its environment into more population is a generalization of what is taking place in the substructure of a biological system.

A population and its environment are the two reactants in a biological reaction and metabolism is the reaction. The product of a biological reaction is more population. The first distinction between a chemical reaction and a biological reaction is the process of metabolism. Metabolism is an extended series of chemical reactions that catabolize the variety of resources provided by an environment, then anabolizes them into population. Unlike a chemical reaction in which each reactant is a volume of homogenous atoms or molecules, an environment is rarely composed of homogenous materials. However, the population reactant in a biological reaction is composed of homogeneous individuals that have the same structure and function.

The more important difference between a chemical reaction and a biological reaction is that a chemical reaction usually produces a product, or products, which are structurally and functionally different from either of the initial reactants. In contrast, a biological reaction is autocatalytic. The product of a biological reaction is always more population, and the product population is both structurally and functionally homogeneous to the individuals in the initial reacting population. This of course causes a biological reaction to have a continual feedback loop between the product of the reaction and the population input into the reaction which makes the biological reaction autocatalytic.

As with all autocatalytic reactions, the extent and duration of the reaction is limited only by the non-autocatalytic reactant. For a biological reaction between a population and its environment, this means that the production of population is limited only by the capacity of its environment and that the reaction will continue to expand until that capacity is reached. This means that environment limits population.

Environmental capacity is subject to Liebig's law of the minimum. This law states that growth is not dictated by the total resources available but by the scarcest resource, which acts the limiting factor to growth. In a biological reaction Liebig's law of the minimum determines environmental capacity and limits the growth of the population.

A biological reaction has a reaction surface just as in a chemical reaction. It is the area in which population and its environment meet and where the population accesses the resource of its environment. By being the area where resources are supplied to the population, a reaction surface causes the distribution of a population to conform the distribution of its environment. This, along with the feedback loop between the output population and input population of a biological reaction, causes the density and distribution of the population to conform to the density and distribution of its environment. Thus, an environment not only limits the volume of population but directs where individuals are located within the population.

The idea that replication and reproduction are the biological analog to a chemical reaction offers an explanation to August Weismann's observation that germ line cells are often sequestered from somatic cells. The environment for the molecular populations composing a cell is carried within the cell's cytoplasm, and specifically for the genetic population comprising the cell's nuclear DNA, the environment is contained within the nucleoplasm. Were the nuclear DNA of germ line cells to contact somatic cytoplasm, it would likely begin to react. Development would then follow whatever surface those environments supplied. However, maternal germ line cells are not only sequestered from somatic cells but also stored in a form much like most viruses awaiting a host, which is DNA encapsulated in an inert lipid capsid devoid cytoplasm. Just as laboratory chemicals are stored in inert glass containers to keep them from reacting or degrading, nature stores her reproductive genetic populations in inert capsules until ready for use.

From a physical perspective, development arises from the biological reaction between a population and its environment. When a population reacts with its environment, more population is produced, creating an autocatalytic feedback loop that forces the population to expand to the capacity of its environment. The reaction surface between the population and its environment expands in relation to the distribution and density of the environment, causing the distribution of a population to conform to the distribution of its environment. A population's environment therefore determines the form of the population's distribution.

LAW OF CONSERVTION OF MASS

In a biological reaction, the volumes of the reactants and the amount of waste produced in converting them can cause the volume of the product produced in the reaction to be quite variable. This makes it rather difficult to mathematically model a biological reaction as it is done with chemical reactions. In chemical reactions, specific volumes of reactants always produce specific volumes of product. However, the imprecision inherent in a biological reaction does not relieve it from being subject to the laws of thermodynamics and specifically the law of conservation of mass and energy.

The law of conservation of mass and energy, which maintains that mass can neither be created nor destroyed but only converted, dictates that no greater volume of population can be produced than there are resources to produce it. The abundance of offspring produced by some species appears to be caused by the fact that in most of nature the initial cost, in resources, of initiating reproduction is many times less that the cost of sustaining an individual throughout its lifespan. Hence, many more are produced than survive a full lifespan. Quite obviously, a population's consumption of resources cannot exceed the volume of resources provided by its environment, but neither can the volume of population exceed the volume of resources supplied by its environment. If an environment suddenly diminishes, the volume of population must also diminish. In regard to reproduction, the total number of individuals being reproduced will never at any time exceed a volume that is supported at that time by the volume of environment. A population's environment therefore sets a limit as to the maximum volume of population at any given time; and as the environment fluctuates, so does that limit.

In general, environmental resources are produced continually. Sunlight, water, heat, inorganic materials, organisms, other organic material, and many other things continually enter an environment and become resources for a population. Once available to the population, these resources can be consumed and utilized by the population. They may degrade if not used or they may exit the environment in some other manner. This continual movement of resources into and out of an environment creates a flow of resources through the environment that sustains the population. An environment is a flow of resources, and a population is a flow individuals supported by the reaction between the two. The flow of population cannot exceed the flow of environment. In other words, the law of conservation of mass prescribes that at every instance of the present, there are no more individuals existing than there have been resources to produce and sustain them.

The law of conservation of mass prevents the volume of a population from exceeding that which is sustained at any point in time by the flow of resources through its environment; the feedback loop, between the product population and the input population in the biological reaction, causes a population to always expand toward the carrying capacity of its environment; and the reaction surface between a population and its environment causes the density in the distribution of the population to reflect the density in the distribution of its environmental resources. These three relationships cause the volume of a population to always fluctuate toward the capacity of its environment, but not exceed it. In a dynamical sense, this means that the flow of population moves toward equilibrium with the flow of environment, so that the volume of population always moves toward equilibrium with the capacity of its environment. There appears to be no exception to this rule.

These relationships also cause the distribution of a population, through its circulation and flow, to adjust to the distribution of its environment. Over time, fluctuations in the circulation and flow of a population will mimic the circulation and flow of its environmental resources, causing the distribution of the population to conform to the distribution of its environment. This relationship, that the distribution of a population conforms to the distribution of its environment, is universal to all populations.

The form of a population is its distribution, and an environment, by determining the distribution of its population, determines the form of the population. The importance of this is that biological organisms are systems composed of populations, and the form of each population within a system contributes to the overall form of the system. The distribution of an environment by determining the distribution of its population, indirectly contributes to the form of the system. In other words, environment determines form.

GROWTH & MATURITY

An organism's growth is determined by the growth of its compositional populations. These populations cannot grow beyond the capacity of their environments and their environments are limited by the position that the organism occupies, as well as by the ability of the organism to convert its positional resources into environments for its compositional populations. Consequently, an organism's position, by limiting the growth of the organism's compositional populations, limits the growth of the organism.

The environments within a system are produced by the functioning of the system, which converts the resources of its position into environments for each of its populations, and the growth of each population is limited by its environment. A position is then not only the total of direct resources available to a system, but also the total of indirect resources available to be converted into environments by and for the populations that compose the system.

In a developing system, the system is expanding or growing toward equilibrium with capacity of its position. This growth is caused by the populations within the system expanding toward the capacity of their environments. As each population within the system expands, its activity of converting resources expands along with it, and so too does the output of that activity which contributes to the environments of the other populations within the system. The populations of a system will continue to expand relative to one another until the use of resources by the system reaches equilibrium with the capacity of the resources flowing into its position. Thus, the growth of a system is limited by its position.

The populations that compose a system are mutually dependent on one another to produce and provide environments. This interdependence not only causes them to function as a system but also forces them to grow relative to one another during development. A population cannot outgrow its environment, and its environment is produced by the populations that form its system, so the growth of each population within a system is regulated by the growth of the other populations in that system. The growth of the entire system is limited by the volume of resources provided by its position. Although numerous factors can affect an organism's actual growth, its position sets the upper limit of its potential growth. That growth is also subject to Liebig's law of the minimum.

Once the volume of a developing population reaches the capacity its environment, its activity, and the output associated with it, ceases to expand. When the environments produced within a system no longer expand and the populations occupying them have expanded to capacity, the growth of the system has reached maturity. At maturity, all the populations that form a system have expanded to the capacity of their environments, and the system they form has expanded to the capacity of its position. Maturity occurs when the consumption of resources by a system reaches equilibrium with the capacity of the resources provided by its position. For most organisms, growth ceases or diminishes at maturity.

For the populations that compose a system, maturity is the point where each population ceases its initial expansion and begins to fluctuate in equilibrium with the capacity of its environment. These populations have then expanded their distributions to the limit of their environments, giving the system its adult, or mature, form. The system is then also at equilibrium with its position, and from that point on it attempts to maintain that equilibrium. Maturity stabilizes both the structure of a system and the metabolic pathways that were established during its development.

While the mature populations that form an ecosystem can fluctuate quite readily, through reproduction and mortality, with the fluctuations in their environment, the highly integrated populations that form an organism have much less ability to do so. Organisms often undergo structural changes at maturity that prevent them from resuming their developmental pattern of growth once maturity has been achieved. Hence, if an organism's position substantially expands after it has reached maturity, the organism has only a minimal ability to consume the increased resources of its position. Moreover, if its position substantially declines after maturity, the organism again has only a minimal ability to reduce its consumption of resources. An organism cannot simply shrink its size in response to a diminishing position because its sequence of development is not perfectly reversible.

The size of an organism at maturity sets a narrow range of consumption necessary to sustain the organism, and the organism will then attempt to maintain its level of consumption throughout its maturity within this range. If the volume of resources availability at its position falls below this range, the organism will perish and if the volume expands beyond this range, the organism will reproduce.

REPRODUCTION

Maturity occurs when an individual's consumption of resources reaches equilibrium with the capacity of its position, or for a population when it reaches equilibrium with the capacity of its environment. Maturity infers the ability to reproduce, and while populations do not reproduce, individual do, so we tend not to think of populations as mature. In general, a population can be in only one of two states, either expanding toward equilibrium with its environment which is its developmental stage, or in equilibrium with its environment which would be its mature stage. If we define maturity as the point at which equilibrium is reached, then a population does mature, and when an expanding population reaches equilibrium with the capacity of its environment, it may then be considered mature. Once maturity is attained, a population will then attempt to maintain a dynamic equilibrium in which the volume of population fluctuates with the capacity of its environment.

A population expands and contracts to its environment through flow. Flow is simply the movement of individuals into and out of a population through reproduction and mortality. Varying the flow of individuals allows a population to fluctuate along with the fluctuations that occur in its environment. The flow of a population always attempts to increase toward equilibrium with the flow of its environmental resources. This is because the expansion of a population is limited by its consumption of resources which cannot exceed the volume of resources provided by its environment; and because the autocatalytic reaction between a population and its environment always produces population. Thus, a population always attempts to increase toward equilibrium, even in a declining environment. However, regardless of this, a declining environment will always reduce a population because a population can never expand beyond equilibrium. These two countervailing forces always cause a mature population to fluctuate in equilibrium with the flow of its environment.

A population has only a few options with which to respond to fluctuations in its environmental capacity. It may increase its consumption of resources in response to an expanding environment by increasing reproduction, decreasing mortality, or by increasing the amount that each individual consumes. Conversely, it may decrease its consumption of resources in response to a declining environment by reducing reproduction, increasing mortality, or by decreasing the amount that each individual consumes.

A declining environment manipulates the number of individuals in a population first by reducing individual consumption and then by eliminating individuals through mortality. Perishing eliminates the individual, and with it, its position. A decline in resources causes an increase in the mortality rate and a decrease in the reproduction rate so that the mortality rate exceeds the reproduction rate. The increase in mortality occurs in both mature and immature individuals and is due to the inability of the environment to maintain the level of resources necessary to support the existing number of positions. The decrease in reproduction results from a decline in excess resources available to those individuals that do reach maturity. Over time, the number of individuals in the population declines. The opposite occurs in increasing environments, both individual consumption and reproduction increase, causing an increase in population. Increasing resources sustain more positions, provide excess resources for reproduction, and sustain individuals for longer durations.

Any change in the flow of a population is a result of the action of the individuals in the population responding to changes in their positions. An increasing environment increases the volume of resources at its positions, while a decreasing environment reduces the volume. However, the volume of resources at any position within an environment may change independently of the fluctuations in the environment, and the individual alters its consumption of resources only in response to variations in its position. Moreover, an individual alters its consumption in response to its position somewhat differently than a population alters its consumption in response to its environment. The way in which an individual responds depends on whether or not it has reached maturity. An immature individual may consume less or perish in response to a diminishing position while it may grow and consume more in response to an increasing position. However, a mature individual may also respond to an increasing position by reproducing.

Immature individuals tend to increase their level of consumption as they grow toward maturity. The cost in volume of resources to initiate reproduction is often considerably less than the cost of maintaining an individual to maturity. As an individual grows, the volume of resources it needs to sustain itself usually increases until it reaches maturity. This increase is accommodated in a stable environment by the natural mortality occurring within the population. As individuals perish, their positions are eliminated, and those resources are incorporated into other positions. However, in an increasing environment, all positions tend to increase which can allow immature individuals to extend their period of growth before reaching maturity. The greater size produced by the extended growth increases the range of consumption necessary to sustain the individual at maturity.

A mature individual has a narrow range in which it can adjust its own consumption, but if its volume of positional resources fluctuates outside this range, then the individual must either perish or reproduce. Mature individuals have a limited ability to grow and therefore respond to excess positional resources by reproducing. To reproduce, an individual must acquire resources in excess of that necessary to sustain itself. Those excess resources are gained from other positions when the occupants of those positions perish or through an increase in the environment. Reproducing subdivides the reproducing individual's position into more positions, thereby reducing the availability of excess resources at its position to within its normal range of consumption. In an increasing environment more mature individuals encounter excess resources at their position, causing the reproduction rate to increase. Conversely, in decreasing environments, fewer mature individuals encounter excess resources at their position, causing the reproduction rate to decline.

Because the initial cost of reproduction in material and energy is usually quite low, it appears that while reproduction requires an increase in position, it does not require a substantial increase beyond that necessary to maintain the mature individual. A minimal increase in position is sufficient to induce reproduction and to allow a population to maintain itself at the capacity of its environment.

Reproduction and mortality rates within a population tend to correlate directly with the waxing and waning of environments, indicating that an environment through position determines the number of individuals in a population. However, environment can also affect population number by shifting the per capita consumption of its population through phyletic size increase or decrease.

PHYLETIC SIZE INCREASE & DECREASE

A population most commonly changes its consumption and use of resources by altering the number of individuals through reproduction and mortality. A less common method by which the volume of consumption conforms to the capacity of an environment occurs occasionally when an environment causes an adjustment in the population's per capita consumption by adjusting individual size through a phyletic increase or decrease in stature.

The minimum level of consumption necessary to support an individual tends to increase throughout development and the attainment of maturity sets a range of consumption that a mature individual must maintain. This range sets a minimum level of subsistence below which the mature individual will perish and above which, it will reproduce. Because the subsistence level of consumption is not set until maturity, the state of an individual's environment and position during its development can cause the minimum level of subsistence to increase or decrease upon reaching maturity.

In an increasing environment and position, an immature individual will tend to increase its consumption of resources and extend its period of development leading to the attainment of a greater size and greater minimum level of consumption at maturity. Conversely, in a decreasing environment and position, the individual's consumption and growth will be retarded leading to a smaller size and lower minimum level of consumption being attained at its maturity.

Given a sufficient number of generations, a continual expansion or contraction of an environment may cause a phyletic change in the stature of a population's individuals. A change in the average size of the individuals in a population will cause a corresponding adjustment in the per capita consumption of resources. The result being a population of larger individuals consuming more resources per individual, or of smaller individuals consuming fewer resources per individual. A phyletic change in stature adapts a population's consumption to its environment by changing the volume of resources necessary to support each individual rather than by a changing the number of individuals consuming.

This can be described by using an analogy. Consider that an environment is a room and that the individuals of a population are boxes of the same size, then a particular number of boxes are needed to fill the space of a particular size room. If the size of the room is expanded, then more boxes of the original size are needed to fill the room. If the size of the room is reduced, then fewer boxes of the original size can fit into the room. This is the normal state of the relationship between an environment and its population. However, there are situations in which a change in the size of the room could lead to a change in the size of the boxes that fill it, so that the size of the box expands or contracts along with the size of the room. Changing the size of the box is analogous to a phyletic increase or decrease in the average size of the individuals in a population. A phyletic change in size causes a corresponding change in the volume of resources that are necessary to sustain each individual.

Now consider how this would apply in expanding and declining environments. In a declining environment, which often occurs prior to a population going extinct, each generation of the population would encounter fewer resources than their predecessors and thus fewer resources would be available per position. As the population regenerates, individuals in each future generation would find the volume of resources provided by their position in the environment to be less than that of their predecessors. At each position, this would retard the point of equilibrium, between volume of consumption and volume of resources, where an individual reaches maturity. The duration of time it takes for the individual to reach maturity would then also be reduced along with the individual's duration of growth and its ultimate stature at maturity. Over time, this would cause phyletic dwarfing within a population.

However, for this to occur it seems that the rate of decline in the environment would need to continue over many generations and at a rate sufficiently slower than the generation rate of the population so as not to cause immediate extinction of the population. Such phyletic dwarfing of a population should be somewhat rare, should always be associated with environmental decline and often lead to extinction.

The converse of this phenomenon should also occur in expanding environments. In this scenario each generation in an expanding environment would find their positions to have a greater volume of resources available than in the previous generation. The duration of time necessary to reach maturity would be extended, allowing additional growth in the developing individuals, and causing the average size reached at maturity to increase. However, for an expanding environment to cause a phyletic size increase, the rate at which the environment expands would need to be sufficiently in excess of the regeneration rate of the population and continue over several generations. It appears then that such a scenario of sufficiently rapid and sustained environmental expansion may be so rare as to likely never occur. Yet evolution is replete with phyletic increases in size.

There are two possibilities for this. First, successive environmental expansions may cause small increases within single generations that accumulate over many generations. Second, it is possible that a rapid phyletic increase in size could occur when a small incipient population encounters a vast and unoccupied environment, such as often occurs with invasive species. In this scenario, the environment is not necessarily changing or expanding but rather a small population, relative to the size of the environment, has adapted to the environment and is attempting to fill it. Each generation of the population would encounter positions well in excess of that to which they can expand, causing the point of maturity to be extended with each generation until the population eventually reaches equilibrium with its environment. The result of this would be very rapid population growth and, over several generations, an increase in the average size of the individual. Phyletic size increase, therefore, always indicates that a population is expanding.

The importance of position as it relates to the development of form is that it determines the number of individuals in a population. The rule that a population conforms to is environment implies that a population must adjust its resource use toward equilibrium with the availability of resources provided by its environment. It can do this in one of two ways, either by changing the number of individuals consuming resources or by changing the volume of resources that each individual consumes. Each method has different causes and outcomes. Changing the number of individuals involves changing the rate of reproduction and mortality. Changing the volume of resources that each individual consumes requires a change in the average size of the individual. By directing which method is employed, the environment, through position, determines the number of individuals within a population.

Of the four variables that determine form, location, number, and variety are determined either directly or indirectly by environment. The distribution of an environment determines the location of the individuals in a population. Environment and position determine the number of individuals in a population. The number and the location of the individuals in a population determine the population's distribution, and distribution gives the population its form. The form of a system is the combined forms of its compositional populations, and changes in the distribution of one or more populations within a system can alter the form of a system to produce a new variety. The fourth variable of form, timing, is also determined by environment.

THE TIMINING OF DEVELOPMENTAL EVENTS

Within a developing system, the point at which equilibrium is reached between the use of resources and the capacity of those resources provides both a timing and a triggering mechanism for any changes that occur in the course of the system's development. The point of equilibrium, at which the increasing use of a resource meets the capacity of that resource, both initiates the change and determines the time at which it will take place during development. At the point of equilibrium, the use of a resource shifts from increasing at an increasing rate to fluctuating with the capacity of that resource. This alters the course of development for the individuals, populations, or systems that are reliant on those resources. Single resources, environments, and positions all have a capacity to support development and the course of development changes at the point at which that capacity becomes fully utilized.

During the development of a system, each of its compositional populations is expanding, and each is increasing its output of converted resources. This enables the expansion of the other populations within the system. Once a population reaches equilibrium with its environment, its output of converted resources stabilizes. This can have a stabilizing effect on any other population in the system that is supported by those resources. The process of stabilization in the development of a system begins to occur as the consumption of resources by the system reaches equilibrium with the capacity of its position.

As described previously, a system reaches maturity when it has expanded to the capacity of its position. At maturity, it can no longer expand as it did during its growth phase and any further development of the system comes in the form of reproduction rather than growth. The attainment of equilibrium between the use of resources by the system and the supply of resources by its position triggers this change in development.

Within a developing system, each developing population expands to the capacity of its environment and tends to do so at an increasing rate. This is caused by the expanding reaction surface that is produced by the autocatalytic reaction between the population and its environment. During a population's development the course of its expansion is directed by the reaction surface. Once equilibrium is attained the reaction surface no longer expands but becomes equivalent with the distribution of the environment, and any further development of the population then directly follows the fluctuations in its environment. This can alter the course of development of the system in which the population resides.

A system is an integrated set of populations in which each population assists in converting positional resources into the environments that support the other populations in the system. In a set of populations that form a system, each population is supported by resources produced by one or more of the other populations in the system. The metabolic pathways of a system are the processes that produce and distribute resources to the various populations. During the development of the system, each population expands toward equilibrium with its environment and the volume of resources each population produces for the system expands along with it. Once a population ceases to expand, the volume of resources it produces also ceases to expand, and those populations whose expansion is dependent on those resources will either cease to expand or change the variety of resources they utilize for their expansion. Populations only cease to expand once they reach the capacity of their environment.

While a population cannot expand beyond the limit of its environment, an expanding population in a diverse environment can outstrip the capacity of a single resource, causing the continued expansion of the population beyond that capacity to be supported by alternate resources. Equilibrium with the capacity of a resource causes the course of any further development of the population to conform to the distribution of the alternative resource. This can affect the form of the system produced with that popualiton.

At the point of equilibrium expansion either ceases and the onset of maturity begins, or expansion is continued through use of alternate resources. The volumes of single resources, environments, positions, and alternative resources are all determined by environment, which means that environment, either directly or indirectly, determines the timing of developmental events throughout the levels of organization in biological structure. Thus, as with the other elements of form, the timing of developmental events is also environmentally determined.

FORM AS A PATTERN OF MOTION

As noted previously, it is the movement of matter that distinguishes individuality. When a set of objects moves relative to one another, as do the units of form that compose a system, it allows the whole of a system's units of form to be distinguished as a single entity, even as they move. Looking closely at a cell, one will see that it is composed of molecules that are continually moving. As with a cell, it is true for any biological system that its units of form are in continual motion, either through circulation or flow. Molecules continually move within cells, cells move within organisms, and organisms move within the biosystem. All biological structures and their units of form are in continual motion, this includes all individuals, populations, and systems as well as the environments and positions they inhabit. This movement normally follows a repeating pattern.

Although we perceive an organism's form to be fixed, it is continually changing over time due to the movement of its units of form. The shape of an individual, a population, or a system, if viewed over time is a pattern of motion in the circulation and flow of its units of form. This movement is usually ignored because it takes place at a scale either too small to be seen, or too slowly to be noticed. The aging of a person is a patten of movement in the organization of its cells that takes place so slowly that it is barely noticed from one day to the next.

Most, if not all, biological motion consists of short repetitive movements, such as movement back and forth or the repeated forming and breaking of the same kind of chemical bond. These motions are very common in cells, where one such motion instigates another, creating a sequence that forms a longer pattern of movement. The Krebs cycle and protein synthesis are examples of sequential patterns of movement that occur within cells. These pattens of motion are repeated, and often further sequenced into even longer and more complex series, which become the cell's metabolism. Metabolism requires movement in the cell's units of form and this movement follows a pattern that is repeated throughout a cell's lifetime. Patterns of molecular movement produce cells, patterns of cellular movement produce various organisms, and moving organisms create ecosystems. These patterns of movement, even highly complex ones, are very repetitive.

Form develops as an extended series of short repetitive patterns of motion combined into longer patterns, and compounded at greater levels of organization. Metabolism, cell division, embryonic development, growth, aging, and even evolution are all examples of repetitive patterns of motion in the circulation and flow of units of form. In fact, all organisms develop as some complex aggregated series of simple patterns of motion. The more elaborate and complex the pattern of motion, the more time it takes to unfold. The development of an offspring is a repeat of the pattern of movement that took place in the molecules and cells which produced the parent. Repetition of movement is not exact, but it is close enough to produce the same form with incredible accuracy. This repetitive movement produces consistency in the development of individual form throughout a population and throughout time.

A pattern of movement is the sequence in which units of form move through a series of locations and the arrangement of those locations in space relative to one another. The more a pattern repeats a sequence and the more that sequence follows the same spatial locations, the more repetitive and stable it is. Alternatively, the less repeatable it is, in sequence and in space, the more variable and less stable it is. A biological form is the pattern of motion of its compositional parts, rather than just the fixed location of those parts. Biological form is created and sustained by the movement of its units of form. The pattern in which units of form move will continue for some time, and the duration of that time represents the life span of the form.

How is it that biological life can so accurately produce repetitive patterns of motion? The answer is that units of form are confined and regulated by the space in which they move, and individual units of form move efficiently through that space. The distribution of an environment sets a physical boundary that confines the circulation and flow of a population while position confines the movement of the individual. Individual units of form move efficiently, and that movement is regulated by both the structure of their position and of their environment.

Consider for example an hourglass, which is a simple physical system that both repetitively and accurately creates a form. When turned over, an hourglass allows its sand to fall through the aperture between its upper and lower globes. The sand forms a shape in the lower globe that has a hemispherical bottom and a conical top. The sand will repeatedly create this same shape each time the hourglass is turned over, and it will do so indefinitely. Gravity provides the energy which gives the sand its motion and the globes of the hourglass, which confine the sand, direct the sand grains to form its particular shape in the lower globe.

The shape of hourglass along with gravity creates a physical environment that distributes sand in a manner that produces a particular form. The sand requires no genetic program to construct this shape. The design of the hourglass produces a pattern of motion in sand grains that causes them to create a distinct and reproducible form each time the hourglass is turned over. Similarly, an environment confines the movement of its population, and to the extent that both environments and populations remain the same, the repeated filling of an environment with produce the same distribution in the population.

In the same manner as the hourglass, metabolic and morphogenetic pathways limit the motion of individual units of form in a biological system causing them to move and act in a particular manner. While it is the force of gravity that moves sand grains, the movement of biological molecules is affected by such things as gravity, light, heat, chemical gradients, chemical affinities, catalysts, and inhibitors. Biological development is a much more complicated series of processes than the hourglass, yet it follows the same fundamental principle of organization, and the result is the same. Biological form is a pattern of motion, and boundaries restrict that motion just as the shape of the hourglass restricts the motion of sand grains falling through its aperture. An hourglass can be flipped indefinitely, and so long as its structure is maintained, the sand will produce the same form an infinite number of times. In comparison, consider the equally nearly infinite number of times that biological reproduction can faithfully replicate the same form. A biological form will be replicated so long as the environment that produces it is sustained or reproduced.

Repetitiveness in biological motion is the frequency in which units of form reproduce the same distribution. In the hourglass, sand grains end up in relatively the same position each time it is turned over, creating a repetitive form. However, the same unit of form does not necessarily have to repeatedly move into the same position nor even repeat its previous motion, to reproduce the same form. Rather it is the flow of the entire population moving through the same series of positions that makes a pattern of motion repetitive. Repetition in biological patterns of motion does not rely so much on the same unit of form moving in a way that repeatedly returns it to the same position, although this may occur in some biological cycles, but rather as one unit of form departs its position, a similar unit of form takes its place. So that both the circulation and flow of the population follows the same pathway.

The flow of a population moves repetitively because similar objects respond in the same way to the same stimulus, and similar environments exert the same stimulus. A response to a stimulus is movement, so similar objects respond to similar environments by moving in similar manners. This is the source of repetitive motion in biology. Patterns of movement become repetitive when similar individuals move through the same environment. Moreover, the more stable an environment is through time and space, the more stable will be the pattern of motion it produces in its population. This is because patterns of movement become stable when similar stimuli are applied consistently to similar units of form over time. The more stable an environment, the more consistent the stimulus it applies to the population over time.

Let us consider another physical example of repetitive motion, that of water molecules flowing down a river. The form of a river usually looks much the same from day to day except for changes in the water level. Individual water molecules are the units of form in a river, and when energy is applied to water molecules in the form of gravity, they can do little more than move toward the force of gravity unless blocked by some structure.

The shape of a river is created by billions of water molecules constantly flowing down the river's channel. Rapids, pools, eddies, and bends in the river give it its recognizable features that remain much the same over time. Yet we know the water molecules in a river are constantly moving and that the river we see today is not that same river we saw yesterday because the water molecules that constituted the river yesterday have moved on, and new ones have taken their place. In aggregate the flow of water molecules follows the same path with later flowing water molecules following much the same path as earlier flowing ones, while individually no two water molecules ever follow the exact same path. This is because the riverbed creates an environment that directs the flow of water molecules, and the population of water molecule affects the movement of each individual water molecule. A river's form is the repetitive motion of its water molecules confined by the space in which they move.

A very similar thing occurs through the flow of cells in our own bodies. As dying cells are replaced by new cells, a flow of cells is created, which overtime makes us a new person composed of different cells than we were before, and yet we are the same person as we were before. Like the river we look much the same as we did before, but we are not composed of the same units of form as we were. The difference between biological flow and the flow of water in a reiver is that a population cannot exceed the capacity of its environment. It would be as if a river always flowed at the capacity of the riverbed but never overflowed. Cells, like water molecules in river, follow a pattern of flow that gives shape to the organism. Populations circulate and flow in repetitive patterns of motion that give form to their system, and a system's lifespan is the duration to which this pattern of motion is sustained.

5. REGULATING MOTION

FEEDBACK LOOPS

The biosystem has so far evolved to four levels of organization represented by the molecular, the cellular, the metazoan and the biosystem levels, in ascending order of complexity. Molecules, cells, multicellular organisms, and eusocial colonies are the units of form that compose these levels. Each greater level of organization is produced by the units of form that make up all its sublevels.

Organisms develop at each level of organization as individual units of structure form populations that conform to their environment. The replicative reaction between a population and its environment is autocatalytic, which causes feedback loops to arise within each level of biological organization. Transitional systems, which act as an individual at a greater level of organization and as a system at a sublevel of organization, transfer the effect of the feedback loops between levels of organization. These feedback loops cause developing systems to have a significant effect on regulating their own development. They also give a system the appearance of self-organization because the environments produced within a developing system affect the development of the system. However, development at each level of organization within a system is ultimately determined by environments and positions that are external to the developing system.

Within each level of organization, the development of form proceeds through two feedback loops, one that loops between system and population and the other loops between population and individual. The form of a system develops through the first feedback loop (Diagram #1). Environments, derived from the distribution of resources within a system, determine the distribution of, and therefore the form of, each population within the system. The form of each population contributes to the form of the system. The form of the system then determines how it converts the resources of its position and distributes them as environments. As those resources available to a system at its position are processed through this feedback loop, and the loop is repeated over time, the system develops. The volume and variety of resources available to support a system's development is determined by the system's position.

It is important to note that Diagram #1 represents only one population in a system. Systems are composed of multiple populations so there is a similar feedback loop for each population within a system. The arrow can be read as "determines," that from which the arrow points has a determinative effect on that to which the arrow points.

DIAGRAM 1:

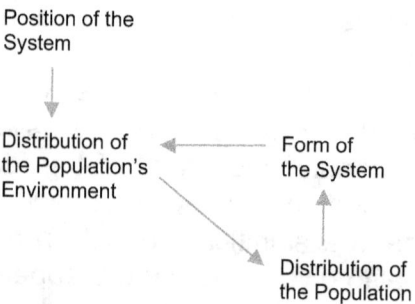

A second feedback loop (Diagram #2) occurs between the population and the individual. In this loop both the distribution of the environment, and of the population, determine the limit of resources available to the individual at its position. Position determines whether and when an individual will reproduce, thereby determining the number of individuals that will be contributed to the distribution of the population. In this loop, environment and the distribution of the population determine position; position, by determining reproduction, determines the number of individuals; and the number of individuals affects the distribution of the population. As this feedback loop repeats itself, and resources are processed through it, the population develops. The limit to which a population can develop is determined by its environment. This feedback loop exists for each individual in the population.

DIAGRAM 2:

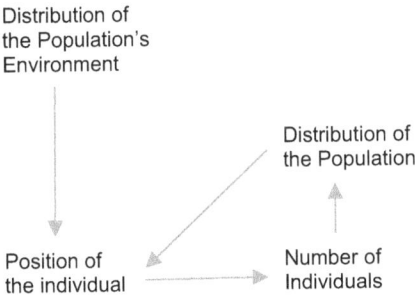

The feedback loops shown in Diagram #1 and Diagram #2 are connected by the distribution of the environment and by the distribution of the population because both participate in each loop (Diagram #3). This allows the position of the system to affect the distribution of each population composing the system and the distribution of each population to affect the form of the system. In this loop, the position of the system and the form of the system determine the distribution of the environment for each population in the system. By its effect on the content and distribution of the environment, the position of the system also affects the distribution of the population. The distribution of both the environment, and of the population, determine the position of each individual. Position determines the number of individuals, which affects the distribution of the population and in turn affects the form of the system. This combined set of loops is ultimately limited by the position of the system.

The position of the system determines the resources available to be converted by the system into environments. It thereby has a determinative effect on the environment of each population within the system. It is also important to note that the form of the system has no effect on its position, so there is no "determinative arrow" shown between the form of the system and the position of the system. The positional resources of the system are converted by the system and distributed as environments to the populations composing it. Hence, there is an arrow from the system's position and from its form to the distribution of the population's environment. Diagram #3 combines Diagrams #1 and #2, and it represents a single level of organization.

DIAGRAM 3:

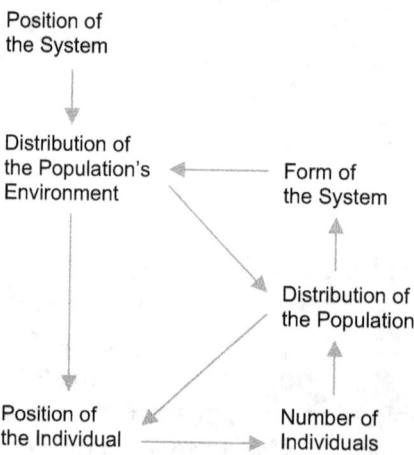

These loops, which operate at each level of organization, are extended between levels of organization through the transition that occurs when a system, produced at a lower level, functions as an individual at the next greater level of organization (Diagram #4).

In Diagram #4, all the levels of organization are shown, and each structure is labeled according to its level of organization. Cells, multicellular organisms, and eusocial colonies are the transitional structures between levels of organization that act as both individual and system. Diagram #4 illustrates how the biosystem affects a single gene residing within an organism that is part of a eusocial colony.

Diagram #4 shows the determinative relationships between structures, within and between levels of organization. It also illustrates how a change in the environment at one level organization can affect development at sublevels of organization. The determinative relationships are generally top down with the greater levels of organization setting environmental and positional boundaries that limit development in their sub-levels of organization.

Causal feedback loops permeate the entire hierarchy of biological organization. Beginning with biological molecules, this loop passes through successively greater levels of structure constructing cells, organisms, and eusocial colonies. Positional resources are distilled down through these levels of structure and substructure, being modified and subdivided into the environments that direct the formation of biological structure at the various levels of organization. Through this process biological structures are produced and conglomerated into the trophic systems that form the biosystem. The biosystem influences the form of all the organisms produced within it. That influence is more direct for those populations that exist in environments produced by the biosystem, but it is further removed for those populations that exist within intermediate levels of organization. For example, to diagram the causal relationships in the development of a single celled organism such as a bacterium, we would eliminate the metazoan and multicellular levels from diagram #4 and connect the molecular level to the biosystem level. The biosystem's influence on the structure and form of a bacterium is much greater than its influence on that of a cell residing within a eusocial organism. The more levels of organization between an individual and the biosystem the less the biosystem's direct influence on the individual. Greater levels of

DIAGRAM 4:

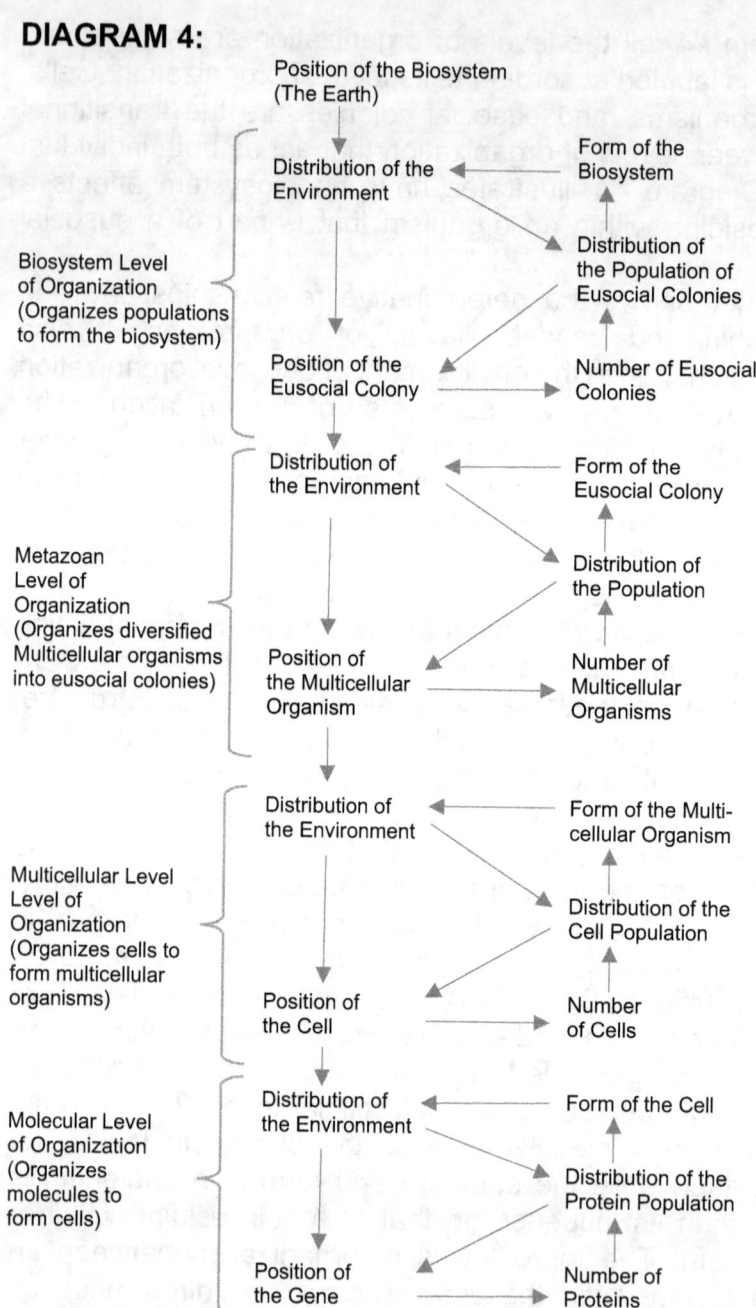

organization appear to produce increased environmental stability within their sub-levels of organization.

The biosystem functions as a single metabolic process that converts energy and biomass throughout its structure in support of itself. It is composed of molecular organisms, cellular organisms, multicellular organisms, and eusocial colonies. All of which are units of form in the biosystem, and which form the populations that together create the form of the biosystem (Diagram #5). The position of each organism that exists in an environment produced within the biosystem level of organization is determined by this feedback loop, and it illustrates the causal relationships that produce changes in the biosystem over time.

DIAGRAM 5:

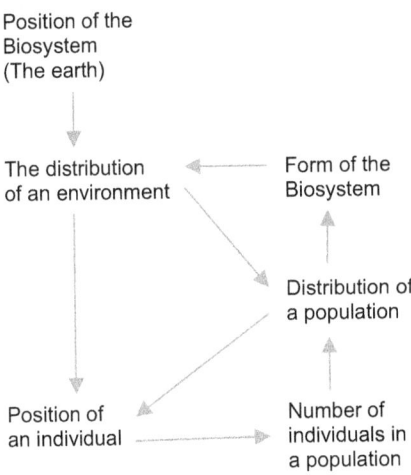

The feedback loops in the above diagrams show that the environments produced within a system are derived from the resources available to the system at its position. The structure of a system – its form – converts the resources of its position and distributes them as environments to the populations that compose the system. Each of the populations composing a system conforms its spatial organization to the distribution of its environmental resources, and the distribution of each population contributes to the structure and form of the system. This means that the structure of the system determines the distribution of each environment produced within it, the distribution of each environment determines the distribution of its population, and the distribution of these populations compose the structure of the system. In other words, the form of a system is produced by populations that conform to environments produced by the system's own form (see Diagram #1). The development of each system therefore responds to the resources available at its position, however, a position does not respond to the developing system within it, making position the determinative constraint on development.

A position is the subset of an environment's resources that are available to a single system, and it is the total of resources that the system can convert into the environments that will support its compositional populations. The form of the system determines how its compositional populations convert the resources of its position and distributes them as environments. These environments both support and constrain the system's populations. The action of the populations within a system, by conforming their consumption of resources and their distributions to their environments, cause the system to adjust its consumption of resources and its form to conform to its position. In this way every system attempts to maintain equilibrium with the constraints of its position, and an organism is a system that functions in this manner.

Diagram #4 can be viewed as a single catabolic pathway from the biosystem down to the individual gene and an anabolic pathway from the gene up to the biosystem. Energy and matter are processed down the diagram while increasingly complex units of form are constructed up the diagram. Cells, multicellular organisms, and eusocial colonies are transitional structures that link the levels of organization by acting as systems at a lower level of organization and as individual units of structure at the next greater level of organization.

These feedback loops may also be considered a passive form of communication between a system and its units of structure. A system, through its production and distribution of environments, communicates to its populations how each population should be distributed. In turn, the distribution of a population divides its environment into positions, and the resources available at each position communicates to its individual how it can react with its position. By converting the resources available at its position, an individual then develops as the correct variety and reproduces when appropriate. This allows information on the volume and variety of resources available at the system's position to be communicated to its units of form, and through the units of form to direct the development of the system to conform to its position.

There appear to be no linear processes in biology, but rather biological systems are composed of a multitude of connected non-linear processes. This increases the efficiency at which individuals conform to their position, and at which populations conform to their environments. Each cycling of a non-linear process within a system moves the system closer to equilibrium with its position, and over time a system is constantly adjusting toward that equilibrium.

EFFICIENCY

All the functions of an organism's metabolism support its subsistence, development, growth, and reproduction by reorganizing matter. An organism's function is to convert the resources of its environment into more of itself. Its structure determines which resources it can convert and how it converts those resources. Converting resources requires movement, either by the organism moving toward its resources or by its metabolism in converting those resources. Every biological unit of structure, whether a molecule, a cell, or an organism, is in continual motion and that motion is directed toward the conversion of its environmental resources. In its most basic meaning, conversion refers to the reorganization of matter through metabolism, and this requires movement. For this reason, all biological motion is equivalent to, and essentially synonymous with, the conversion of resources, and it is through the conversion of resources that biological form is organized.

Energy is required to sustain the motion that produces biological organization, and it is infused into biological systems by the absorption of sunlight through photosynthesis and transferred within and between systems through symbiosis and predation. The energy infused into a system is constantly moving through the system, and in doing so it is producing organization. That organization is represented by the volume of biomass contained within the system.

Energy is expended by a system in converting resources and it is retained within a system by being transferred between its units of form. Both the transfer and the expense of energy produces organization through the pattern of movement that units of form follow in transferring and expending energy. The less energy a system expends in converting the resources necessary to sustain itself, the more efficient it is, because a given amount of energy then sustains greater movement which produces more organization. The more an amount of energy is transferred between the units of form within a system before it dissipates out of the system, the more organization it produces and the more efficient the system is. The more efficient the system, the longer and the larger it can be sustained. Efficiency may then be defined as the amount of energy retained within a system as it moves through it. By this definition, biological systems evolve toward greater efficiency. This is nicely demonstrated in the evolution of the biosystem which over the course of its history has consistently evolved toward increasing the amount of biomass sustained within the biosystem.

Energy is expended through motion, and it is dissipated out of a system as the heat which that motion produces. The amount of energy that flows into a system is equivalent to that which flows out of the system. So, defining efficiency as the amount of energy retained within a system as it moves through, merely allows more efficient systems to be identified as those which are larger or in which energy takes longer to move through. Moreover, identifying the relative efficiency of a system does not explain why a system develops and evolves toward greater efficiency. In other words, identifying what is efficient does not necessarily describe what causes efficiency.

Efficiency may be more specifically defined as the property of motion in which movement is in the direction of least resistance given the immediate circumstances.* When energy is expended by a unit of form it will move and that movement will always be in the direction that provides the least resistance, which means that units of form always move efficiently. This does not mean that a pattern of development necessarily follows the most efficient pathway to some end, but rather it proceeds in the direction of least resistance given the present circumstances at each point in time. Development is reactive, not teleological. However, feedback loops continually move in repeating patterns of development toward greater efficiency.

To produce repeating patterns of development, something must direct the movement of units of form within their environment and that something is efficient movement, in other words movement in the direction of least resistance. All biological movement is efficient which causes the conversion of resources and the development it produces to be directed toward efficient processes. This movement is efficient because energy is not expended unnecessarily overcoming resistance. In the constantly moving process of converting resources, molecules, cells, and organisms all move in the direction that provides the least resistance, thereby producing the greatest organization for the least amount of energy expended.

Units of biological form are in constant motion and a unit of form can move in a myriad of different directions and react with a variety of resources, creating multiple developmental potentials. For patterns of development to be repeatable, individual units of form must have a means of choosing between alternative motions. This is where the property of efficiency comes into play. Efficiency gives direction to the movement of individual units of form, causing them

---

* Efficiency, as defined here and in the glossary, appears to be a fundamental law of motion, but it has never been stated as such even though the concept has been expressed in various ways throughout the past millennium.

to move along the path of least resistance within a varied environment. This establishes developmental pathways, for without efficient movement, individual units could move randomly, impairing their ability to produce and maintain stable patterns of motion. This becomes primarily important when environmental space is substantial and diverse, and the population within it is sparse, as it is during developmental periods. The repetitive patterns of motion in diverse and variable environments are both a necessary and a fundamental characteristic of living matter. Repeating patterns of motion could not be produced if units of form could move randomly within their environment.

Molecular Efficiency

In the standard model of the atom, electrons seek the lowest energy state, or the ground state, which is the most stable form of the atom. In this state the energy contained within the atom will sustain the atom's structure for the longest possible duration, making it the most efficient configuration. The structure of the atom is made up of sub-atomic particles that are in constant motion, and this motion requires energy. The ground state conserves energy by expending the least energy to maintain the atom's structure. Atoms bond to form molecules by sharing electrons, and they share electrons in the most stable electron configuration possible. Molecules bond in the same manner, so we may then say that the structure of a molecule is efficient because it maintains its structure by expending the least energy necessary to do so, and thereby retains energy within its structure for a long as possible. Molecules are the fundamental units of biological structure, and they move efficiently. We may therefore expect those structures built of efficient molecules to also function efficiently.

Both chemical reactions and biological conversion tend to be efficient in that they attempt to expend the least energy for the present circumstances. Elements and molecules will bond preferentially with those elements and molecules that require the least energy to form ionic and covalent bonds and produce the lowest energy state. In this way chemical reactions are efficient in their use of energy. The physical principal of efficiency is carried over into biological organization where both biological molecules and organisms preferentially replicate themselves using those resources that require the least energy to convert.

Physics defines energy as the ability to do work, and work is the ability of an object to move itself or another object. We recognize energy being expended by the movement it produces. In biology, work is performed by metabolism which at the molecular level converts resources into units of biological structure. The concept of efficiency presumes that biological molecules have a propensity to bond with one another that is determined by their atomic and molecular structure, their distance apart, and probably their orientation to one another. This propensity produces a hierarchy of efficient interaction among the molecules in any given distribution, and it is such hierarchies that produce the sequential action necessary for repeating patterns of motion to arise.

Increasing Efficiency

By conserving energy in the process of converting resources, efficiency causes patterns of development to skew toward the production of more efficient systems. An efficient system is one that sustains the greatest amount of biomass for the energy infused into it, given its structural constraints. Environment, position, and an individual's ability to convert resources are the structural constraints that limit efficiency. Overcoming structural constraints through adaptation, often leads to greater efficiency in the biosystem.

Biological systems develop and evolve toward greater efficiency. If efficiency is defined as the amount of movement produced from an amount of energy, then greater efficiency produces greater movement from a given amount of energy. In biology, movement is equivalent to the conversion of resources, therefore greater movement from an amount of energy produces and sustains a greater volume of biomass. Efficiency can then be determined by the volume of biomass sustained with in a system. More efficient systems sustain greater biomass per unit of energy than less efficient systems. By this metric, the biosystem evolves toward greater efficiency.

The biosystem becomes more efficient by producing and sustaining an increasing volume of biomass, and throughout evolutionary history, the biomass sustained within the biosystem has increased. The methods through which evolution has increased the biomass of the biosystem fall into three general categories: increasing the area of the biosystem, building trophic structures, and increasing the size of organisms. Around 80% of the earth's biomass is contained in plants, and nearly all of that contained in terrestrial plants. The expansion of plant life from marine to terrestrial habitats produced the earth's greatest increase in biomass. As autotrophs this probably had more to do with access to greater amounts of sunlight and less with the efficient use of sunlight by various species. However, for the biosystem, as a system, it is a more efficient use of the energy that flows to its position. By adapting plant life to terrestrial habitats, the biosystem was able to increase its position within the biosphere by accessing the sunlight that otherwise fell on barren terrain.

The emergence of trophic systems through predation enabled animal life to recycle some of the energy contained in plant life to build additional layers of biomass within the biosystem. Although small in comparison to the biomass of plant life, this is an evolutionary increase in the biomass of the biosystem and an increase in the efficient use of the energy that flows into it.

Increasing the size of organisms seems to play a role in almost every evolutionary increase in efficiency. In trophic systems, size differences allow larger animals to prey on smaller animals The evolution of increasingly larger species expands the structure of a trophic system by adding additional conversion layers that reuse the energy entering the system. The additional layers of increasingly larger predators increases the biomass sustained by a trophic system without increasing the energy flowing into the system.

Transferring energy through symbiotic relationships has enabled the emergence of greater levels of organization, which in size are orders of magnitude larger than the individual units that compose them. Multicellular organization has increased the size and spatial range of autotrophs enabling them to capture more of the energy entering the biosystem and provide a larger base to support the heterotrophs that recirculate that energy.

The principle of efficiency is derived from the propensity of units of form to conserve energy by moving in the direction of least resistance. Over time, this can increase the energy retained within a system. Efficiency causes the greatest amount of biological organization to be produced from the energy absorbed into a system, given the structural constraints of the system. History demonstrates that evolution progresses intermittently toward greater size, greater complexity, and a greater volume of biomass sustained within the biosystem. By evolving toward greater biomass carrying capacity, the biosystem becomes more efficient by sustaining more movement per unit of energy flowing through it.

Efficient Movement of a Population

A single unit of form does not have to follow any specific pathway within a developmental process, rather what is important is that the population follow a particular pathway and that the positions in a pathway are filled by the correct type of unit of form. This occurs because individuals continually move along the path of least resistance. This creates what often appears to be the random movement of individuals within the population, while the population as a whole follows a particular pathway within the system. A population will then develop in a predictable pattern in a given environment. Thus, the structure of the environment, the distribution of the population, and the direction of individual movement all participate in a pattern of development.

Consider again water flowing down a river, which is a physical example of this. No two water molecules follow the exact same pathway down the river, and yet for the river to look the same from moment to moment, all the pathways along which water may flow down the river must contain a water molecule. What is occurring in the water molecules is simply that no two individuals can occupy the same space, so the occupation of space creates resistance that is not present in unoccupied space. As one water molecule moves, space is made available that is unresistant to occupation by another water molecule. So as the water flows downstream, each water molecule follows a random path, but every possible pathway is filled with a molecule, causing the combined flow of water molecules to follow a specific pathway. The movement of the population of water molecules is confined by its environment which is the course of the riverbed. This same principal holds true in the movement of biological populations within a developmental process.

A different description of this same example would be to suppose that we track the path of a single water molecule as it moves down the course of a river, and then track the paths of all the other water molecules traveling down the course of the river. We would find that no single water molecule followed the exact same path as the first water molecule, but whenever and wherever we looked at the path of the first water molecule there would be a water molecule on that pathway. Moreover, we could arbitrarily designate a pathway down the course of the river, and we would find that there is always a water molecule filling every position in that pathway. The eddies, rapids, and bends in the river are all consistently produced by the flow of water molecules, and yet no water molecules follow the exact same path down the river, nor has any intention of moving along a path that would create these features. These features are created by the riverbed constraining the movement of water and each water molecule moving along its own path of least resistance.

Another example is that of the hourglass mentioned earlier. The grains of sand in the hourglass are a population of similar individuals that respond similarly to their environmental influences. No two sand grains follow the same path as they fall through the aperture between the globes of the hourglass, eventually coming to rest in the lower globe. And yet the population of sand always produces the same shape in the bottom globe each time the hourglass is turned over.

The process of development is similar to water molecules flowing down the course of a river or sand in an hourglass. It does not matter that a specific molecule or cell follows a specific pathway throughout the course of development, but rather it is important that throughout the course of development a population follows a particular pattern of movement. So long as every individual moves in the direction of least resistance, a moving population of similar individuals with the same functional abilities and subject to the same environmental stimuli will deliver individuals to the correct place and time over the course of development.

An individual's movement is both confined by its environment and by the distribution of the population, and that area in which it is confined to moving is its position. Because individuals move efficiently, rather than randomly, an environment through its distribution of resources can direct the movement of the individual within its position. Efficiency creates the consistency in individual movement that causes a population to follow a consistent pattern of movement within a given environment. This is especially necessary during development when environments are diverse and sparsely populated, and a specific sequence in the pattern of movement is essential to the organization of biological structure.

Efficient Resources

Efficiency is determined by the amount of energy expended in producing a given volume of biological structure. The greater the efficiency of the individual, population or system, the less energy is needed to sustain it. However, the resources of an environment can also have greater or lesser efficiency, and the efficiency of a resource can change over time. Efficient resources are those that cost an individual the least energy to acquire and to convert, and not all environmental resources are created equally in this respect. A population will consume those resources that are most accessible and the easiest to convert first, leaving those that are less accessible and harder to convert until the more efficient resources become less accessible.

It does not matter whether energy is expended in the acquisition of a resource or the conversion of the resource. The total amount of energy expended by the individual in accessing and converting a resource determines its efficiency. For example, competition for over utilized resources expends energy, making the more efficient resources less efficient for those individuals less able to compete for them. This also increases the relative efficiency of lesser efficient resources where there is less competition. When individuals move efficiently, energy is not expended unnecessarily overcoming resistance.

The efficiency of a resource may depend on its location relative to the individual. Resources that are nearer the individual may be more efficient than those farther away. Less energy is expended moving toward close resources than in moving toward more distant resources. The efficiency of a resource may then change along with the movement of the individual or the resource. Thus, the efficiency of a resource is not the same for all individuals in a population.

HOMEOSTASIS

Homeostasis is the dynamic equilibrium that a system maintains with its position. It is accomplished through a system's compositional populations adjusting to fluctuations in their environments. In mathematics, non-linear systems are attuned to their initial conditions, but biological systems are not because development is normalized at each level of organization by the limit of the position in which the system resides. This normalization leads to homeostasis. Homeostasis is the continual flux of a system toward equilibrium with its position, and all biological systems are homeostatic.

In biology, homeostatic systems can be described as either primarily symbiotic or primarily predatory. Symbiosis and predation being the two extremes in a range that characterizes the conversion of resources. The molecules in cells, the cells in multi-cellular organisms, and the individuals in eusocial colonies generally function symbiotically while ecosystems are generally predatory. Each population within the system performs some function in the conversion of resources that provides environments to the other populations within the system. In symbiotic systems, both resource conversion and distribution are generally carried out through circulation – the movement of individuals within their environment. In contrast, the ecosystems that compose the biosystem convert resources primarily through flow by means of predation, where one population consumes the members of one or more other populations within the system.

In predatory systems, the conversion of resources requires the destruction of the prey by the predator, which affects the rate of flow in the prey population. In a normally functioning ecosystem, predation has little effect on the volume of the prey population. This is because the total biomass of a population is dependent on the volume of its environmental resources. The removal of individuals through predation, frees up environmental resources to be used by the remaining population. These excess resources tend to be invested in reproduction. This increases the rate of flow within the prey population to the capacity of its environment. The volume of individuals flowing into the population becomes roughly equivalent to that exiting through predation. Predation, rather than reducing the total biomass of the prey population, increases its rate of flow to that which can be sustained by the environment.

Populations adjust to their environments, and the external elements that may affect their environments, by increasing or decreasing their activity. That activity is the conversion of resources through circulation and flow. As environments fluctuate, symbiotic systems tend to increase or decrease circulation to accommodate those fluctuations. In multicellular organisms, most of this activity is carried out at the molecular level within and between cells, as most cellular populations within a multicellular organism do not circulate. By adjusting molecular circulation, a multicellular system can regulate its physiological processes and body chemistry to maintain its stable and balanced functioning. This is quite the opposite in trophic systems, where animal populations tend to circulate quite readily. In these systems, populations adjust primarily through increasing or decreasing flow – the production and demise of individuals. While conversion through predation seems to be the easiest to evolve, symbiosis appears to be the most efficient, and it is present in all levels of organization.

Whether symbiotic or predatory, the environment of any one population within a system is dependent on the activity of the other populations within that system. Thus, the populations within a system regulate each other through the provision of environmental resources to one another. This creates a feedback loop in which each population, by adjusting to the fluctuations in its own environment, causes fluctuations in other environments within the system, to which the other populations must adjust. This feedback loop directs the process of homeostasis.

The production of environments by a system's populations is ultimately dependent on the resources provided to the system by its position. The system's position is then the limiting factor on any fluctuations taking place within the system. The system, as the whole of its populations, will always attempt to fluctuate toward equilibrium with its position. This is the basis of homeostasis in all biological systems.

Development is itself a homeostatic process that integrates the populations of a system as the system expands toward equilibrium with its position. In maturity, homeostasis is the system's means of adjusting to fluctuations in its position, but homeostasis is a limited mechanism for doing so. Extreme changes in the volume or type of resources, or the loss of crucial resources at a position can severely damage or destroy the individual occupying that position.

While homeostasis, and the feedback loops that facilitate it, cause a system to move toward equilibrium with its position, equilibrium is never actually attained because each movement of a unit of structure, or of the resources that support it, moves the point of equilibrium. A system will then never reach perfect equilibrium with its position but will continually move toward it. This causes homeostasis to be continually ongoing process of dynamic equilibrium between a system and its position.

COMMUNICATION

A system's ability to regulate itself through homeostasis makes it apparent that biological systems have the ability to communicate with their units of form. This is generally assumed to be genetic in origin because it initially arises in development and is thought to direct the organization of form. However, the existence of chimeric organisms suggests that it is a characteristic of the structural organization of biological systems.

A chimeric organism develops from two distinct genomes and each genome builds that part of the organism complimentary to the other. The idea of genetic determinism suggest that a genome is programmed to build an entire organism, yet somehow in a chimeric organism two distinct genomes know to build only that part of the organism not built by the other genome. This suggests that during development information, in some manner, is being communicated between the two genomes.

Inherent in the structural organization of biological systems is a means of passive communication between the system and its units of form. The development of a system not only produces individuals and populations but also develops diverse environments and subdivides those environments into positions. A unit of form can utilize only those resources that it has access to at its position, and its development and functioning will be constrained and directed by those resources. A system thereby directs the development and functioning of each of its units of form by distributing resources to their positions. A cell in a chimeric organism, or in any other organism, then does not need to know where it is in relation to the other cells in the organism, it simply needs to be able to function properly in its position. If it does so, it will do its necessary part in support of the system.

The two distinct genomes of chimera are directed to build complimentary parts of the organism, not because there is direct communication between the two genomes, but rather because the resources allocated to each position in the developing organism direct development at that position. The two different genomes have the same functional abilities, and so will build similar cell types in similar positions. Those cells will also distribute themselves similarly in similar environments and function similarly in similar positions. The system directs the development of each cell, regardless of its genome, by providing the correct volume and variety or resources to the correct position.

A system communicates to the individual through the system's allocation of resources to the individual's position, while the individual communicates back to its population through its use of resources, and back to the system through the output that its conversion of resources adds to the system. The allocation of resources to a position, by limiting the functioning of the individual occupying that position, tells the individual what it can produce. The functioning of the individual adds structure and resources back to the system, which tells the system its form and function.

The populations within a system communicate among each other through the structure of the system. We see the effect of this in homeostasis which is a result of the communication between a population and its system. The function of a system is to convert and distribute resources to its compositional populations, and those populations form the structure of the system. This creates a communication feedback loop between the functioning of a system and the functioning of its populations. This feedback loop regulates the system and causes its populations to function relative to one another.

The most basic level of communication is between an individual and its population through the individual's position. An individual acts according to the resources available to it at its position, and the availability of resources is determined by its environment and the use of resources by all other members of its population. Because of this there is a higher degree of communication by the population and its environment directed toward the individual and very little by the individual directed toward its population. The effect of the population is entirely directed toward the individual, while effect of the individual is spread out among the entirety of the population. The intensity felt by the population from the action of the individual is negligible in comparison to the intensity felt by the individual from the action of the population.

Communication between the individual and its system travels through the population, which weakens the intensity of the individual's effect on the system but does not eliminate it. Communication between levels of organization travels through the transitional individual that acts as a system in the sub-level of organization and as an individual at the greater level of organization. Because communication can travel between levels of organization, a biological molecule in the cell of an organism does then communicate with the biosystem, and the biosystem with it, and thus there is then a pathway of communication between every living unit of form in the biosystem. However, the intensity of such communication is normally so weak that it may be considered as non-existent.

6. DETERMINANTS OF DIFFERENTIATED FORM

VARIATION & ADAPTATION

A variation is a difference in form that does not change an individual's ability to function. The function of an individual is to convert resources, and a change in function allows an individual to convert its resources more efficiently or to convert resources that it was previously unable to convert.

Function is an inseparable aspect of form. The design of an organism evolves for the intended purpose and function of converting the resources of its environment into more of itself, either through growth and subsistence or through reproduction. An organism's manner of functioning is its ability to convert resources which determines the kinds of resources it can convert and how it converts them. A particular form enables the conversion of a particular set of resources in a particular manner.

Individual forms can vary to some greater or lesser degree and yet function similarly. Phenotypes vary among the individuals of a population because their units of form are distributed differently. For example, one person is different from another person because the cell populations that compose their bodies are distributed in slightly different configurations, yet they function similarly because the characteristics that distinguish the two forms do not confer any meaningful difference in their ability to convert the direct resources that support them.

An organism is a system composed of populations and each population has a distribution that contributes to the form of the organism. A variation is caused by a difference in the distribution of one or more of the populations that compose the organism. A change in the distribution of a compositional population alters the form of the organism, and a sufficient change in the distribution of enough populations within an organism may not only alter the form of the organism but also change the way in which it functions. Such a change in form is an adaptation, and an adaptation produces a new variety.

A variation may then be defined as a difference in the distribution of a compositional population that alters the form of its system, but which is not sufficient to change the functioning of the system. Variations may accumulate over generations to become adaptations, and an adaptation may be defined as a difference in the distribution of a compositional population that is sufficient to produce a change in the functioning of its system. An adaptation increases the efficiency in which a form converts its resources. Changes in form that increase the efficiency of a form may also allow it to convert resources that it was otherwise unable to convert, thereby increasing the diversity and expanse of its environment.

By this definition of adaptation, cellular differentiation in embryogenesis and the diversification of species are both a result of adaptations. Cells that differentiate during embryonic development have different forms and functions. An organism that has evolved a new form and function is a new species. Evolution and development both produce adaptations but at different levels of organization, which suggests that they are a result of the same process being carried out at different levels of organization.

Adaptation does not necessarily require a change in the variety of the individual units of form. A change in the distribution of the same varieties is sufficient to cause a change in function. For example, all mammals are produced from the same approximately three hundred different types of cells. The difference between a deer and an elk is not in the variety of cells with which they are composed, but rather in the distribution of those varieties. Deer and elk evolved from a common ancestor into distinct species because over several generations their cellular units of form evolved different distributions.

From a structural point of view, evolution requires only that there be a change in the distribution of the units of form that is sufficient to change the functioning of the organism. Much of the phylogenic diversification that has occurred throughout evolutionary history may have been accomplished merely through changes in cellular distribution, it did not necessarily require novel genes or novel cell types. For one species of mammal to evolve into another requires only that the existing variety of cells be distributed in a different manner. It generally does not require the acquisition of new genes or the evolution of new cell types. In fact, not only do mammals share the same variety of cell types, but genetic research is beginning to suggest that the same set of functional genes are common to all mammals. Mammal species differ from each other in their form and function, not because their cellular units of form are of different types or that they necessarily have different genes, but because the variety of units of form they have in common are distributed in different configurations. In this view a mouse genome has the capacity to produce the form of a man. The reason it does not is because the developmental process of the mouse does not distribute cells in that manner.

An adaptation is a change in the functioning of a system, and adaptations presumedly arise through the accumulation of variations. A variation is difference in the distribution of one or more of the populations that form a system. The distribution of a population conforms to its environment, and a population's environment is produced by its system. Thus, the structure of the system produces the environments that determine the distribution of the populations which produce the structure of the system. Where then does variation get introduced into this feedback loop of circular causation?

A population's environment is produced by the system in which it inhabits, and the system produces that environment from the resources of its position, so changes in position become the source of variation in form. A change in the content of a system's position can change the content and distribution of the environments the system produces for its compositional populations. This, in turn, can alter the distribution of those populations and consequently the form of the system. However, this usually occurs only during the developmental phase of a system and not in mature systems.

Development distributes a system's populations efficiently which conforms the developing system to its position and allows differences in position between the parent and its offspring to be accommodated in the offspring's development. Development allows flexibility in the organization of compositional populations, whereas the organization of a mature population is confined by the capacity of its environment. At maturity, the structure of the system has conformed to its position, and the causal feedback loop between the structure of a system and the distribution of its compositional populations causes resistance to any later changes in its position. When environmental change occurs, it is not the mature organism that alters its form to conform to its changed position, but rather its offspring that develops in the new position. This means that variation and adaptation are effects produced by development.

The principle that a variation in form can be achieved by altering the spatial organization of a compositional population within the substructure of a system, occurs at each level of organization. For example, changing the distribution of proteins can change the form of a cell, or changing the distribution of cells can change the form of the organism. It is becoming clear from the genomic mapping studies carried out on an increasing variety of mammals, that Individuals representing the diverse species of the same clade are all composed of the same variety of cells and those cells are built from the same functional genes. What appears to vary is the karyotype and the phenotype. In other words, the difference between a horse and a donkey is not in the kinds of cells with which they are made, or in the genes which produce those cells, but rather in the distribution of those cells and the distribution of the genes within the DNA of those cells. This is in accordance with the structural point of view that differences in distribution are the source of variances in biological form.

A change in the distribution of a system's populations causes variation, not in the form of the individuals that compose those populations, but rather in the form of the system that is composed of those populations. Because the phenotype of a system is produced by the distribution of its compositional populations, and a compositional population's distribution is determined by its environment produced by its system from the system's positional resources, the form of the system is indirectly determined by its position and any variation in its form is caused by changes in its position. While position does not directly determine the individual's form, it does provide the pathway through which the individual's environment determines its form.

Of the parameters of form – variety, number, location, and timing – all are in some way determined by environment and position. Variety is indirectly determined by position, the number of individuals is a result of reproduction which is determined by position, location is determined by the distribution of environmental resources, and timing is triggered by equilibrium between a system and its position and by equilibrium between a population and its environment. Variation, adaptation, and the varieties that arise from them are the result of changes occuring in one or more of the environments produced within the multi-level structure of biological organization.

MOLECULAR SYNTHESIS & DIFFERENTIATION

Molecular Organization

A cell is a molecular system composed of a large variety of molecular populations. Like all biological systems, its form is derived from the variety, volume, and distribution of its molecular units of form. However, a cell is unlike higher-level biological systems in that its units of form are not individual biological systems, but rather individual molecules.

A cell's molecular units of form include lipid and carbohydrate molecules that are synthesized by the cell from the resources it ingests. Lipids form much of the cell's structure and membranes, while carbohydrates generally provide energy and energy storage for the cell. In addition to these, a cell also produces two basic types of nucleotide base molecules: proteins and nucleic acids. Nucleic acids include DNA and RNA. Proteins come in several different kinds including structural, storage, contractile, transport, antibodies, enzymes, and hormonal proteins. Proteins do the work of the cell; they convert the positional resources of the cell into the molecular environments that within the cell support its molecular populations. The chromosomal genes are the replicating population that initiates the production of the cell's proteins.

It takes several tens of millions of individual proteins to build a cell, but a cell, a human cell for example, may contain only around 12,000 different types of proteins. The different types of proteins represent the variety of molecular populations that make up a cell. About 10,000 of these are structural proteins. Most cells are constructed of roughly the same types of structural proteins, but the concentration may vary significantly among different kinds of cells. In addition to structural proteins, each cell type may contain as many as a few thousand more kinds of proteins that are in some variety and concentration specific to that cell type.

The human genome contains around 20,000 different genes, which appears to provide considerably more variation than its cells use. In addition to this, the cell can produce protein variation in excess of that contained within in its genes. Each gene can produce multiple forms of a protein depending on how it is transcribed, cut, spliced, or excised, and these can have multiple kinds of post-translational modifications. For example, a methyl or phosphate group can be attached, or they may be joined to a lipid or carbohydrate, all of which affect their function. Possibly as many as 100,000 different proteins can be produced from 20,000 genes.

A cell is in fact a microscopic jungle of molecular populations. Assuming a cell is made up of forty million molecules and those include 12,000 different kinds of proteins as well as various lipids and carbohydrates, it is probable that there are many more than 12,000 different molecular populations within a single cell, since a type of protein may have more than one population within a cell. This can happen if individuals of the same type are segregated to different areas of the cell and subject to different micro-environments. Having different environments, these groups of identical proteins would constitute separate populations that may function differently due to their differing environments.

With millions of individual proteins of many thousands of different kinds, a cell is a system composed of a great diversity of molecular populations, each with its own microenvironment within the cell. Like other biological systems, in which the system's positional resources are converted into environments, a cell employs its molecular structure to convert its positional resources into environments that will produce and sustain its molecular populations.

Protein Synthesis

Proteins molecules do the work of the cell by converting the cell's positional resources into environments and distributing those environments to the cell's molecular populations. In relation to other units of form at higher levels of organization, proteins are highly specific in function. It takes several thousand different proteins to produce a functioning cell, but at the multi-cellular level of organization it takes only a few hundred different cell types to form a functioning organism. Mammals for example are composed of around three hundred cell types. At the eusocial level of organization, a functioning colony can be produced with only a handful of different phenotypes. Functional specificity in units of form appears to decrease as the level of complexity increases.

Proteins originate from the genes residing on the chromosomes. In conventional biology, the chromosomes are considered organelles, and as a group they are one part of the reproductive organ of the cell that produces the protein populations of the cell, but in the convention of structural organization, they are a population of individual genes. This is because they have similar structure and function and exist in the same environment. Individually, a gene occupies a position within its population that is determined by the distribution of both the population and its environment. Genetic positions along the chromosomes are mostly immobile. Every gene has a position that is determined by the flow of resources to the chromosome and by its location in relation to the population of genes on the chromosome. A gene's position both affects, and is affected, by the other genes in the population.

The environment of the chromosome is largely composed of free nucleotides circulating within the nucleoplasm, but it also contains many other molecules such as enzymes and other proteins that can catalyze or inhibit transcription or modify a transcribed gene. This genetic environment is produced by the cell's molecular populations that function together as a system to catabolize the cell's positional resources into free nucleotides and other molecules. These molecules are then infused into the nucleus to become the molecular environment of the chromosomal genetic population.

While chromosomes do move, the position of each gene relative to the other genes on its chromosome is stationary, which means that genes circulate only to the extent that their chromosome circulates. A population of genes must then conform to its environment through flow. Flow is produced by the production and demise of individuals within a population; a gene's demise usually comes with the demise of the cell it inhabits but its reproduction occurs in two ways. The first is with the full replication of the chromosomes during mitosis and meiosis, and the second is when it initiates protein synthesis through transcription. In both, the gene is responding to the volume and variety of resources available to it at its position.

The idea that the volume and content of resources in the molecular environment of the gene can have a regulatory effect on the genetic replication is implicated by the effect that position has on reproduction. Because genes are composed of only four nucleotide bases, the volumes of these bases relative to one another probably plays a lesser role in genetic replication than does the volume and distribution of molecular catalysts and inhibitors within the nucleoplasmic environment. Reproduction, as described earlier, occurs when a mature individual encounters resources at its position sufficiently in excess of its range of consumption. Reproducing subdivides the reproducing individual's position into more positions, thereby reducing the availability of excess resources at its position to within its normal range of consumption. Genes do not grow or mature, rather they are molecules that either react or do not react with whatever environment they encounter. For a gene to replicate, it merely needs to be in contact with the correct variety and volume of molecular resources. A very minimal amount of environment may be sufficient to replicate an individual gene. However, it does appear that a substantial excess of environmental resources is necessary to replicate the entire complement of chromosomes. This is apparent in the initial divisions of the embryonic egg cell during embryogenesis in which a series of cellular divisions takes place without any protein synthesis occurring.

The egg cell is a rather large and cytoplasmicly voluminous cell. This suggest that there is some level of nucleoplasmic volume at which below that volume individual genetic replication takes place, and above that volume chromosomal replication takes place, just as there is a volume of positional resources that causes an individual to change from merely subsisting to reproducing. At that volume of nucleoplasmic content, the nucleoplasm shifts from acting as the environment for a population of genes to acting as the position of the entire chromosomal complement. If the volume and content of the nucleoplasmic environment can regulate between a cell's genetic replication and its chromosomal replication, then it likely also regulates the differential replication of genes.

Genetic Replication and Expression

A gene is like all other chemical structures in that it either reacts or does not react with other chemicals. Like other biological units of form, its function is to replicate itself, but unlike units of form that are biological systems, a gene cannot grow. Therefore, its only function is to replicate. Whether or not a gene replicates and how often it replicates is determined by its position in relation to other genes and to the content and flow of its environment.

Single genes are replicated through genetic transcription which copies the gene into messenger RNA (mRNA). Protein synthesis begins with the transcription of gene. Free nucleotides attach to a section of the chromosome that codes for a particular gene, producing an mRNA template of that gene. That template may then undergo post-transcriptional modification before it exits the nucleus and makes its way to the ribosome. In the ribosome, the mRNA template is copied back into DNA by transfer RNA (tRNA) molecules that combine amino acids in the order specified by the mRNA template, this produces a polypeptide chain of amino acids. That polypeptide may then undergo post-translational modification before it condenses into its final structure as a functional protein. While this is an abbreviated description of the sequence, it points out that genetic transcription is only one step in the process of protein production and the only step reliant on the gene, the remainder of the process is carried out independent of the gene. Both post-transcriptional and post-translational modifications are crucially important to producing the functional structure of a protein and neither is controlled by the initial gene.

Gene expression is not necessarily the transcription of a gene, but rather the differential translation of genes. If one gene is expressed more than another, its mRNA template is translated more frequently to produce more proteins, a protein is the expression of a gene. Without producing a protein, a gene has little effect on the structure and function of a cell. Translation is done in the ribosome which translates the mRNA template to produce the primary polypeptide structure that condenses into a protein. A single mRNA template, which is the transcribed copy of the gene, may be translated multiple times by the ribosome. The expression of a gene as a protein is then dependent on both the gene being transcribed into an mRNA template and on the number of times that mRNA template is translated by the ribosome. Both the transcription of the gene and of the functioning of ribosome can affect the extent to which a gene is expressed as a protein.

The more often a gene is transcribed into mRNA, the more likely and often it may be translated by the ribosomes, and the more it is expressed. Similarly, the more times an mRNA template is translated by the ribosome, the more proteins are produced with that gene and the more that gene is expressed. A gene may get transcribed but if it is not translated by the ribosome, then it does not get expressed. Of course, if the gene is not transcribed, then neither can the ribosome contribute to its expression. Thus, the expression of a gene is only partially under genetic control, the ribosome has as much control over the expression of a gene as does the gene itself.

Moreover, a gene cannot be expressed through the synthesis of a protein if the machinery of the cell does not supply a sufficient volume and variety of molecular material to transcribe and translate the gene. Ribonucleotides and deoxyribonucleotides are the basic building blocks of proteins, but there is also a large array of enzymes and other molecules necessary for protein synthesis, and, while the full array of these molecules is found in most cells, their relative volume varies significantly depending on the cell type. The molecular environment of both the chromosomes and the ribosomes is largely maintained by the tightly regulated metabolic pathways of the cell. These pathways involve numerous feedback controls that function to convert the positional resources of the cell into nucleotide and amino acid building blocks and other molecules necessary for protein synthesis. Studies show that disrupting these pools can have a significant effect on both genetic replication and protein synthesis. Environment therefore plays a role in genetic expression as similarly important as the gene and the ribosome.

Gene Position

Through the existence of regulatory genes that switch on and off the transcription of other genes, geneticists have recognized that a particular gene's position in relation to other genes can determine if and when a gene might be transcribed. At this basic level of biological structure, position has a determinative effect on transcription. However, the concept of position goes somewhat further to suggest that all genes have, to some extent, a regulatory effect regardless of how undetectable that effect may be. In this view, the genetic population, through its utilization of molecular resources and its effect on the circulation and flow of those resources, may act as either catalyst or inhibitor on the action of the individual gene depending on each gene's location. The concept of position takes into account the state of the environment. The transcription of a gene is determined not only by the distribution of the population of genes but also by the content and movement of the nucleoplasmic environment, and by the use of that environment by the rest of the gene population.

If genes cannot change their position relative to one another, then a position can be altered only by changing the direct resources that flow to a position. Altering the variety of genes transcribed is necessary for differentiation to take place in a cell lineage. This suggests that differential gene activation depends on the composition of the chromosome's environment, where changing the composition alters the variety of genes transcribed. It appears that it is the gene's environment and not the gene that initiates such a change, which implies that it is the molecular environment generated by the cell and supplied to its nuclear genes that directs transcription.

A gene can only react with its environment, it cannot determine when a protein will be produced, how many will be produced or where proteins will be placed in a cell's structure. A gene's effect is only on the type of protein that is produced. Because of this, the genome's effect on development is in limiting what can be produced, not necessarily in directing what is produced. However, this still has a significant impact on the development of form. Given the stable structure of the genome, it may be that there are few viable developmental pathways regardless of the environment.

The structure and function of a cell is determined by the variety and distribution of the molecular populations that compose it. Most of the functioning of a cell is done by proteins and the synthesis of a protein originates with a gene. The singular function of a gene is to produce a template for the protein building blocks of the cell, but a cell's form and function is determined by how those building blocks are assembled. A gene is somewhat like a brick mold that is used to make bricks; the brick mold can tell us what the brick is going to look like, but it cannot tell us how the brick is going to be used in the structure of a building. Similar to the bricks in a building, a protein may have different functions depending on the type of cell it is in, where it is located in the cell, and on the nature of the other proteins with which it functions.

While genes appear to have little ability to direct development, they are none the less indispensable as templates for the protein building blocks of life. Without a full set of the correct genes, development can go awry, and faulty genes can produce faulty proteins. In our building example, faulty proteins are analogous to having faulty building blocks that might crumble. Additionally, the interaction of genes can produce phenotypic characteristics, just as some building material do – the red clay in red bricks will give a building a red appearance. But a phenotypic characteristic does not indicate that a set of genes directed the development of an organism any more than a set of building materials could direct the construction of a building.

Summary

The form of a cell is the distribution of its molecular parts, many of which are made of lipids and carbohydrates, and both of which are not fabricated by the genome. The functioning machinery that does the work of the cell, and generally produces the cell's form and function, is made of proteins, but the fabrication of a protein is only initiated in the chromosomal genes, not fully produced by them. A cell's form is produced by the molecular populations that compose it, and its structure and variety are determined predominantly by its protein molecules.

If the structure of a cell is considered in relation to the four variables of form, it becomes apparent that very little of the production of a cell's structure is controlled by its genes. Nothing within a gene's structure suggests that it can control the number of proteins produced from it, the location in the cell's structure where the protein will eventually reside, or the timing at which a protein is produced.

The number of each type of protein produced is determined by the ribosomes which can translate a single mRNA template repeatedly into a multiple number of proteins. Protein production is dependent on the cell providing a sufficient flow of aminos acids to the ribosomes with which to translates the gene's mRNA template. Although a protein is reliant on the gene for its primary structure, the number of proteins produced from a template is independent of the gene.

The location where a protein ends up in the structure of a cell is tightly controlled by the chemical affinities of the protein and the physical structure of the cell, which include such things as molecular gradients, membranes, microtubule conduits, and the flow of energy within the cell. The existing structure of a cell directs where newly produced units of form will eventually locate in the cell's structure. In other words, the present structure of the cell directs the formation of its future structure.

The timing of genetic replication is critical especially for that that group of genes that determine cell type, but nothing in the sequence of nucleotides that compose a gene suggests it has control over the timing at which it replicates. A gene is a reactive molecule, and when resources are available to a gene, it will react with them. The functioning of the cell as a molecular system gives it the ability to convert and direct the distribution of resources toward its genome. A gene's position, both in relation to its population and to the circulation and flow of its nucleoplasmic environment, seems to be the most probable agent in control of this variable of form, but the structure of the gene itself does not appear in any way to control the timing at which it is transcribed. These factors cause the timing of developmental events within a cell to be largely independent of its genome.

Of the four variables of form only variety is determined by the gene, and due to the transcriptional and translational modifications that can take place after a gene is transcribed, even this is not fully under genetic control.

CELL DIFFERENTIATION

Differentiation in biology occurs for the same reason it does in chemistry, when two chemicals react, they produce a specific product. If one of the reactants is altered, the product is also altered. In biology the reaction is between a population and its environment, the product of the reaction is more population. If one of the reactants is altered, then it will affect the product just as it does in chemistry.

Cell differentiation during embryonic development is a process that takes place at both the molecular and cellular levels of organization. At the molecular level within a cell, the chromosomes are a molecular population of genes that interacts with its nucleoplasmic environment derived from the egg cell's maternal cytoplasm. The molecules dispersed throughout the nucleoplasm are direct resources for the genetic population contained in the chromosomes. At the cellular level of organization, the developing embryo is a developing cellular system occupying a position, as well as a set of diversifying cell population each having a distinct environment within that system.

Embryonic development begins with a single fertilized egg cell, the zygote, which contains DNA from each parent and cytoplasm derived from the maternal parent during oogenesis. In a chemical sense, the nucleoplasm derived from the maternal parent is the environment and fertilization by the male spermatozoa acts as a catalyst which induces a biological reaction between the genetic population of the chromosomes and its nucleoplasmic environment provided by the maternal parent.

The contents of the maternal cytoplasm play a primary role in controlling the initial development of the egg cell. This cytoplasm contains products of maternal genes deposited into the egg during oogenesis such as mitochondria, histones, methyl molecules, free nucleotides, and amino acid precursors, all of which inhibit or promote the transcription of the egg's chromosomal genes. For example, methyl molecules can inhibit the replication of a gene or a gene promoter by attaching to it, thereby enhancing the transcription of other genes. Maternal histone proteins can bind to DNA causing it to condense, thereby inhibiting the transcription of that DNA.

The effect of the maternal cytoplasmic environment on development has been demonstrated by Jean Brachet and his colleague Henri Alexandre at the Free University of Brussels on hybrid merogones between ascidians. Consider the following results of an experiment in which the "…eggs of species A were cut into two halves (merogony) and the anucleate half was fertilized with sperm of species B (hybridization). The development of such hybrid merogones in ascidians is purely of the maternal, cytoplasmic A type." (Brachet and Alexandre 1986) * In this study, the maternal cytoplasm appears to direct the pattern of development rather than the paternal genome.

Unfertilized egg cells are sequestered from the soma devoid of cytoplasm until ovulation. Prior to ovulation the egg cell exists in much the same state as spermatozoa or a virus, it is simply DNA encapsulated within an inert shell. Maternally derived cytoplasm is infused into the oocyte during ovulation and the environment of the cell's nuclear DNA is derived from that cytoplasm. Because the organism that develops in this experiment is of the maternal cytoplasmic form and not of the paternal genetic form, it appears that development is being controlled by the maternal cytoplasm and not the paternal genes of the nucleus. This suggests that the composition of the egg cell's cytoplasm can alter cellular organization in a developing organism.

In Brachet's experiment we may presume that species A, B and the hybrid, all produced the same variety of cells, but that the maternal cytoplasm altered one or more of the four variables of form at the molecular level of organization or the cellular level of organization or both, to produce the hybrid form A using the paternal genes of species B. This suggests that the anatomy of an organisms is not fully encoded within its genome, and that any genome may have multiple developmental potentials depending on the developmental environment. It also suggests that it is the maternal cytoplasm that initially directs embryonic development while the genome merely responds to its environment. The extent to which the life history of the maternal parent affects the content and structure of the maternal cytoplasm of the oocyte is unknown, yet the avenue to do so certainly exists.

---

* Brachet, Jean, and Henri Alexandre. (1986) *Introduction to Molecular Embryology, 2nd Totally Revised and Enlarged Edition.* Berlin: Springer-Verlag. 1986, p. 112; The assertion of natural selection is that the selection of the phenotype selects for the genotype. In this experiment, not only did the genotype not produce the phenotype it was selected to produce, but it produced the phenotype of a different genotype. This is a direct refutation of a primary tenet of natural selection. A genotype that was 'selected for' through one selection process should not be able to produce a phenotype that develops from a genotype that was 'selected for' through a different selection process.

Embryonic Cell Differentiation

As an egg cell begins to divide, there is initially no differentiation between cells because the extremely high concentration of all the molecular constituents within the cytoplasm of the daughter cells allows full replication of the genome which, in vertebrate embryos, causes the cell to immediately undergo cell division. In fruit flies, and probably other invertebrates, the nucleus undergoes several divisions before the egg cell divides into multiple cells, each with one nucleus. The egg cell undergoes several divisions before protein synthesis takes place, which indicates that there are sufficient molecular resources contained within the cell's cytoplasm and transferred to the nucleus to support several full replications of the chromosomes. It appears that in this initial phase, development is completely under the control of the maternal cytoplasm which supplies environmental resources to the genetic population within the nucleus.

The molecular constituents of a cell's cytoplasm are not evenly distributed, and in an egg cell the contents of the cytoplasm are rather distinctly gradated, giving the egg cell its characteristic animal and vegetable poles. The initial divisions of the egg cell replicate the chromosomal genes identically but divides the heterogeneous contents of the cytoplasm unequally and separates it into the dividing cells. Each division of the propagated cells increases the disparity between them in their cytoplasmic content. So, while the cells may still contain the same variety of molecular constituents, division causes the concentration and distribution of each type of molecular constituent to differ among the propagated cells. Once the volume of the molecular environment contained within a cell is sufficiently reduced by division to the point that it can no longer support immediate replication of the full genome, the cell will then begin partially replicating the genome as individual genes. This initiates the onset of protein synthesis. However, because the previous series of cell divisions have sufficiently altered the cytoplasmic content within each cell, the cells no longer provide similar nucleoplasmic environments to their individual genomes.

Each cell division not only changes the nucleoplasmic environment in the daughter cell, but also alters the position of the chromosomal genes in the daughter cell. The chromosomes are a fixed population of genes, so the position of each gene relative to its population remains the same throughout the cell divisions. However, a gene's position is relative not just to its population but also to its environment. Altering nucleoplasmic environment through cell division can change the resources available to a gene at its position, affecting the expression of the gene. Division of the embryonic cell alters the genetic environment in each of its daughter cells, affecting which, when and how often genes are expressed. Once protein synthesis begins, the identical genomes in the daughter cells are reacting with differing environments, this initiates the onset diversification in their cell lineages.

At the cellular level of organization, the initial unequal cytoplasmic divisions of the egg cell propagate a population of cells in which each individual cell contains a distinct molecular distribution within its cytoplasm. Thus, unlike the homogenous positions that exist in the environment of an adult cell population, the positions of this initial cell population are differentiated which leads to differentiation in each of their cell lineages. Once protein synthesis begins, each cell's genome reacts with a nucleoplasmic environment that is sufficiently altered to initiate a differentiating lineage of daughter cells. Further differentiation of these lineages will continue to be caused by the changing molecular environment in subsequent generations of cells. However, the altering of these molecular environments will be less a result of simple cell division and much more a result of how the developing structure of the embryo alters the environments and positions of its cells. During the initial period before DNA synthesis begins, the maternally derived cytoplasm appears to determine cellular activity. However, this cytoplasm can sustain cell division for only a short time before protein synthesis begins and the emerging embryo must begin absorbing nutrients from its surroundings to continue its growth.

As the position and environment of a cell and its prodigy changes during embryogenesis, the environments a cell produces and supplies to its compositional protein populations also changes. This alters the distribution of its protein populations and, in turn, the form of the cell. It is the environment that determines the distribution of its population, and the distribution of the population that determines the form of the system. Hence, a sufficient environmental change at the molecular level, caused either by cell division or a change in cell position, can alter the distribution of the protein populations that compose a cell, resulting in a change in the form of the developing cell.

At the molecular level of organization, the changes in genetic activity that produce cell differentiation are a result of the changing molecular environment of the genome; and while cell differentiation may in general be seen as a result of altering molecular environments and gene positions, which is certainly the cause of differential genetic activity, differentiation within a cell population is rather more fundamentally a result of the expanding cell population dividing its environment. Cell differentiation is caused by the expanding cell population subdividing its environment into differentiated environments, and each of the subdivided populations then efficiently reacting with its portion of the environment. It is the increase in the efficient conversion of a differentiated environment throughout several generations that causes cellular form to differentiate.

For differentiation to occur, the initial environment must be sufficiently diverse, and each subdivision of the cell population must be able to increase its efficient conversion of the resources in its portion of the environment. In other words, while differentiation is induced by environmental differences, it is caused by the increasingly efficient conversion of resources throughout the progeny of a lineage. This natural process of cell differentiation has been artificially reversed in the laboratory by working out the sequence of molecular alterations that lead to a cell's differentiation. Dedifferentiation and trans-differentiation processes have reversed the natural process of differentiation by propagating a lineage of cells and reversing the sequence of molecular changes in their cytoplasmic content.

Differentiation ceases when no greater efficiency can be achieved. A divided population will not differentiate if it cannot increase the efficiency with which it converts its portion of the environment. Nor will the division of a homogenous environment induce differentiation in a population. The division of a homogenous environment simply produces multiple populations of the same type.

CASTE DIFFERENTIATION

The principle, that environment determines distribution and distribution determines structure, not only describes differentiation in cells but also extends to differentiation in multi-cellular form as well. Consider the differentiation in form between castes within a eusocial insect colony, such as that which occurs in ant, termite, or bee colonies. A eusocial colony is a system composed of phenotypically differentiated populations that are represented by the various castes which make up the colony. The individuals of each caste have a distinct form and function that distinguishes them as a population separate from the other castes in the colony.

The phenotype of an ant caste is produced when specific enzymes are fed to undifferentiated larvae. Positions within a population of undifferentiated larvae vary depending on the type of enzymes the larva receives. Those positions that provide the same enzymes, collectively constitute an environment because they provide a slightly different resource set to their population than do other positions. The individuals that develop in those enzymatically similar positions constitute a distinct population because they develop a form and function that is distinct from other individuals in the eusocial system. Each caste originates in its own distinct environment, has its own distinct form and function, and is a population within the eusocial system.

From a structural point of view, it is the distribution of cells that gives the individual caste member its form, not a genetic difference or a difference in the types of cells that comprise its form, as all individuals in a colony are composed of the same variety of cells and all have the same genome, save the diploid queen which has two sets of the same genes. Since all castes have the same genes and are composed of the same variety of cells, caste differentiation must be a result of the distribution of cells in the individual's form. This suggests that caste differentiation is caused by the cellular environments that organize cells within the developing individual.

Cellular environments differ between different castes by the kinds of enzymes they contain. Enzymes are proteins that catalyze metabolic reactions. Catalysts increase reaction rates by lowering the activation energy, thereby making the reactions they affect more efficient, and enabling an increased volume of those reactions. The enzymatic environment within a cell enables the cell's protein machinery to be either more or less efficient – depending on the type of enzyme – in converting the resources of the cell's environment. As a larva undergoes metamorphosis, this difference in efficiency distorts the distribution of its developing cell populations, causing the development of the differentiated caste. This suggests an explanation for the results of Brachet and Alexandre's experiment with ascidians. It is probable that the differentiated phenotypes in ascidians are produced by molecular differences between the maternal cytoplasmic types that induce differential cell population growth during development.

SUMMARY

In biology, a population responds directly to its environment, while its environment is produced through the structure and functioning of its system. A population reacts with its environment to produce more population and as it does so, the environment determines the distribution of the population. A system's structure is the distributions of its populations, which are determined by their environments supplied by the functioning of the system. The system's function is to convert the resources of its position into environments for its compositional populations. Hence, the distribution of the population determines the structure of the system, and this structure of the system determines the distribution of its population. In this feedback loop the only external impetus is the position of the system. Although it is an indirect route, position limits the environments that a system can produce and thereby directs the development of the system's form.

Biological structure is built from the bottom up, starting at the gene which initiates the synthesis of the proteins that form the cell. While genes do have an ability to regulate each other's expression through position, there is yet no evidence that they can directly regulate their own environment, although they likely do so very indirectly through the feedback loops inherent in biological structure. Nor do genes appear to carry anatomical information, rather as molecules they simply react in specific ways with other molecules. Moreover, altering a gene, or simply having a gene, does not necessarily mean that it will be expressed, or if expressed will produce a specific anatomical character. To put this another way, similar molecular environments can elicit similar developmental patterns from different genomes, but similar genomes will not produce the same developmental pattern in different molecular environments. The first part of this statement is evident in the development of chimeric organisms, and the second part is the essence of what Brachet and Alexandre's experiment with ascidians tells us: that molecular environments regulate both genetic transcription and genetic expression.

Anatomy is produced by the distribution of a population and populations are organized by their environments. Genomic structure appears to provide only a slate of genes from which development can draw. In other words, the genome provides a list of the kinds of proteins a cell can make, but which, when, and how many of each protein a cell produces, is determined by the environment the cell supplies to its genome. A gene can do nothing without its environment, and while the same is true for the gene's environment, the product of a gene must conform to the molecular environment in which it resides. In the point of view of structural organization, a genome conforms to its environment through its genes reacting to the resources available at their positions.

While biological structure is built from the bottom up, the environments which organize structure are distilled from the top down, beginning with the biosystem's division into environments. These environments are then divided into positions and each organism converts its positional resources into environments that support its compositional populations. These environments are then again divided into positions. This process of dividing environment into positions and converting position into environments occurs at each level of organization within an organism, and it distills the resources of the biosystem down into the molecular position of each gene.

Both genes and environment affect the development of form. But because biological structure is built from the gene up, genetic effects are diminished as they are transmitted up through the levels of organization. Similarly, the effect of the biosystem is diminished as it is distilled down through the levels of organization. However, it is not the biosystem that organizes structure but rather the environments produced at each level of organization which are derived from it; and each environment directly organizes the units of form that subsist within it. This is why a single genome may have multiple developmental potentials. It is environment that determines the genome's potential; or more precisely, it is the multiple environments that exit at multiple levels of organization that direct the development that is initiated by a genome.

7. DIVERSIFICATION & STASIS

CHARACTERISTICS OF DEVELOMENT & MATURITY PERIODS

Stasis is commonly defined by paleontologists as the lack of adaptive morphological change in a species throughout an extended period during its existence. This is a rather narrow definition given that a species is a type of unit of form in the structure of the biosystem. When a species is in stasis, nearly every other species in the biosystem is also in stasis, which suggests the biosystem itself is in a period of stasis. Moreover, when a species evolves, it tends to split into multiple species, and it tends to do so at a time when many other species are also diversifying. This suggests that it is not the circumstances of the individual or the species that induces evolution, but rather the circumstances of the biosystem.

If evolution is considered from the point of view of the biosystem, rather than from that of the individual or species, then an evolutionary period is a period in which the biosystem's units of form diversify. All biological systems undergo a period in which their units of form diversify. In cells, multicellular organisms, and eusocial colonies, units of form diversify during development. Applying this to the biosystem, we may assume that when species are diversifying, the biosystem, or an ecosystem within the biosystem, is undergoing a developmental period. Conversely, when the biosystem is in stasis, we may assume it is in a period of maturity, and the stasis in its units of form is a result of that.

Cells, multicellular organisms, eusocial colonies, and the biosystem, all have a lifecycle that can be loosely divided into a developmental period and a maturity period. The two periods have distinct characteristics which are apparent to a greater or lesser extent regardless of the level of organization in which the system occurs. These characteristics generally apply to the compositional populations that form the system rather than to the system is itself, but they are expressed in the developing form and the mature form of the system. The division between a system's developmental and maturity periods also coincides with the division between diversification and stasis in the system's units of form. This suggests that the characteristics present in development cause diversification and those characteristics of maturity produce stasis.

The difference between developmental periods and maturity periods, and therefore the difference between diversification and stasis, is environmental space. Excess environmental space allows population growth, while limited space restricts growth. During development, populations are expanding to fill their environments and systems are expanding to fill their positions. Excess environmental and positional space induces that expansion.

Developmental periods are characterized by rapid population growth, greater circulation, irregular patterns of movement, and diversification. In contrast, maturity periods are characterized by stable populations, less circulation, stable pattens of movement and little or no diversification. During maturity, populations are in equilibrium with the capacity of their environments, and systems are in equilibrium with the capacity of their positions. There is little or no excess space available at maturity.

Development is initiated in environments and positions that contain both an excess of resources and a high diversity of resources. The excess volume of resources provides space that allows greater movement and induces population growth through a high rate of reproduction and low mortality. The diversity of resources induces the expanding population to diversify. Developmental periods tend to be very short relative to the duration of maturity because an expanding system soon fills its position.

The characteristics of development are readily apparent in embryonic development. Rapid cellular reproduction, population growth, differentiation, and an initially very high circulation of cells, all contribute to the changing form of the embryo as it develops.

These characteristics subside into stable patterns as the embryo matures; differentiation ceases, cell reproduction and mortality equalize, and the form of the organism stabilizes.

The same characteristics of development and maturity appear in the biosystem. However, the structure of the biosystem is different than that of its sub-level systems and this causes the biosystem to not have a singular period of development followed by a maturity period that extends until the system expires. Rather, the biosystem periodically undergoes short bursts of development interspersed by long periods of relative stasis.

If the biosystem is viewed in its entirety as a singular organism, its history apparent in the fossil record shows distinct periods of growth, decline, and stasis. These may be designated as evolutionary periods, extinction periods, and periods of stasis; but when we look closely at the record of these periods, we find that they are not nearly as distinct as they appear, what sets them apart is their intensity. There are at least five major extinction events identified in the fossil record, each followed by an evolutionary period that regenerated the biomass lost. However, there are as many other periods of extinction and regeneration as one may wish to find, depending on the intensity of the period one wishes to consider. Periods of extinction, evolution, and stasis all can extend for longer or shorter periods of time, be of greater or lesser intensity, and can encompass greater or lesser portions of the biosystem.

This suggests that extinction, evolution, and stasis are characteristics of biosystem development that carry on continually to some greater or lesser degree, and what is recognized as a period of one or another, is in fact the intensity and breadth of one in relation to the others. This means that even during periods of mass extinction, pockets of stasis may be maintained in some remote area or interstice of the biosystem, and some evolution may even occur. Likewise, low grade evolution and extinction may occur during periods of stasis, and some extinction and stasis may exist in periods of intense evolution. The fossil record, as incomplete as it may be, appears to bear this out, because we find that those adaptations which engender growth in the biosystem normally originate long before the opportunity occurs to fully exploit them. Evolutionary periods tend to exploit adaptations developed and maintained in a few marginal species during the preceding period of stasis. We see this quite clearly in the rise of mammals whose adaptations arose in the time of dinosaurs, but which were unable to be fully exploited until space was created by the extinction of dinosaurs.

Periods of evolution and stasis in the biosystem are analogous to the developmental and maturity periods that occur in its sub-level systems. The difference is the level of organization and in the labels applied. We do not think of embryonic development as embryonic "evolution," or the cellular differentiation that takes place in it, as cellular "evolution." Yet, cellular differentiation and speciation are the same process, caused by the same circumstances, only taking place at different levels of organization. The biosystem's evolutionary periods display the same characteristics as developmental periods in lower-level systems. Environments are diverse and minimally populated, circulation is increased, reproduction is rapid, and populations are expanding and diversifying. At the cellular level of organization, embryonic cell differentiation is a process almost identical to the adaptive radiation of species.

Likewise, stasis is the biosystem undergoing a maturity period. The characteristics of stasis in the biosystem are the same as those of maturity in lower-level systems. Compositional populations are at equilibrium with the capacity of their environments and fluctuate homeostatically with them, diversification is at a minimum, and patterns of movement and pattens of development are repetitious and relatively stable.

Maturity occurs when the populations forming a system reach equilibrium with their environments and the system has expanded to equilibrium with its position. Maturity is an extended period in which the system attempts to maintain equilibrium with its position, and its compositional populations attempt to maintain equilibrium with the capacity of their environments. Maturity periods have minimal excess environmental space; reproduction maintains populations at the capacity of their environments; patterns of circulation are stable, regular, and repetitive; environments are less resource diverse with relatively homogeneous positions throughout; and, although a population may evolve anagenetically, little or no diversification occurs. Systems attain their characteristic form at maturity and remain generally stable throughout this period. Their growth rate is substantially reduced relative to that experienced during development, and a system reproduces only during maturity. The duration of a system's maturity period is usually several times longer than its developmental period because development ends when the system reaches equilibrium with its position, while equilibrium may be sustained indefinitely.

Diversification and stasis are characteristics that identify developmental and maturity periods and distinguish between them. In general, development is a period in which a system grows, and its compositional populations expand and diversify, while maturity is a period of stasis in which a system maintains equilibrium with its position, and its compositional populations maintain equilibrium with their environments. Population expansion implies that there is a substantial excess of environmental space containing the resources to support that expansion. Excess space allows movement, growth, and reproduction, and development is the process of filling that space. As excess space fills, development proceeds to maturity, and the characteristics of development give way to the characteristics of maturity.

ANAGENESIS & CLADOGENESIS

Just as cellular differentiation in embryogenesis and cast differentiation in a eusocial colony are environmentally driven, so is the differentiation of species within the biosystem. In traditional cladistics, speciation may occur either through anagenesis or cladogenesis. Anagenesis is a phyletic transformation within a single lineage, meaning that a species, in whole or in part, evolves into a single new sub-species. Anagenesis occurs much less frequently in the fossil record than cladogenesis with the most common example being phyletic dwarfing, or insular dwarfing. In anagenesis, the new species evolves within the geographic range of its ancestral species and usually toward the end of its duration in the fossil record. In contrast, most of the fossil record exhibits cladogenesis in which an ancestral species splits into multiple daughter species during a transition period so short that it usually goes unrecorded in the fossil record.

From the perspective of structural organization, anagenesis is a change in form throughout a population caused by a change in its environment. In contrast, cladogenesis, sometimes referred to as adaptive radiation, is the diversification of a population into multiple forms caused by the expansion of a population splitting its environment into multiple environments. Both anagenesis and cladogenesis are environmentally induced, and both can cause differentiation at any level of organization. The development of cancer may be an anagenetic process at the cellular level of organization, while cell differentiation in embryogenesis is clearly a cladogenetic process. The primary difference between the two processes is that anagenesis occurs in mature systems, while cladogenesis occurs during development and is a characteristic of the developmental process.

Anagenesis is rare precisely because it is difficult for individuals in mature systems to modify their form in the face of a changing environment. An individual develops to efficiently convert the resources of its position, if its position substantially changes as a result of a changing environment, the form of the individual must also change. Mature biological form is resistant to change because it lacks a means to reconfigure itself. The ability of a mature individual to modify itself to a changing position is limited to its range of homeostasis. However, a form can also be modified by regenerating itself. Reproduction allows the developmental process of the offspring to conform the offspring's form to its position in a changing environment. Changes in form always occur over one or more generations, and because of this, an environmental change that induces anagenesis must be mild enough and slow enough to allow regeneration to accommodate it, otherwise the population will go extinct.

Unlike cladogenetic diversification, which splits a population into multiple forms by modifying segments of the population to different resource variances within its environment, anagenetic evolution modifies an entire population to resource changes occurring throughout its environment, this is an important distinction. Anagenetic evolution is caused by changes in the content and structure of an environment that affect the entire population. In contrast, cladogenesis does not necessarily require a change in the variety of resources contained within the environment, although such changes may occur as development proceeds. Diversification through cladogenesis is the result of the seed population expanding into a diverse and unoccupied environment and splitting that environment into multiple differentiated environments, while anagenesis is the modification of a mature population to accommodate its changing environment.

DIVERSIFICATION

The diversification of a population occurs only in developing systems, and it requires an environment that provides both a diversity of resources that will enable differentiation as well as an excess of resources to induce a rapid expansion of the population. The initial population may be a single individual, as it is in embryonic development, or it may occur concurrently in multiple populations as it does in those populations that survive an extinction event in the biosystem. Diversification is induced by an expanding population splitting its environment into multiple regions while dividing and segregating itself into those regions. It is facilitated by the diversity of the environment's resources and the rapid expansion of the population which alters the efficiency of those resources.

A population cannot move beyond its environment; therefore, diversification must take place through the splitting of environments. The process of differentiation is initiated by the division of an environment caused by its population rapidly expanding into it. This division of the environment divides the population and segregates the subsequent increase of population from the original population, separating them into distinct environments. The individuals in the newly segregated population then, through successive generations of efficient development, conform to the positions in their new environment.

The division of an environment is determined by Liebig's law of the minimum. This law states that growth is determined not by the total resources available but by the limiting resource. A limiting resource is that resource which is proportionately the most scarce in the mix of environmental resources necessary to support a population. A population's growth is determined by the capacity of its environment, which according to Liebig, is determined by the capacity of the limiting resource in that environment. This law can be applied to the growth of any biological structure and is applicable to the growth of an individual or a population. When considered along with the relative efficiency of an environment's resources, Liebig's law determines both the timing and sequence in which a developing population diversifies.

Different resources within an environment have different efficiencies, and in developmental environments those efficiencies can change. The efficiency of a resource can be diminished or eliminated for some individuals when their access to that resource is restricted. Resources are always a finite quantity, and the use of a resource by a group of individuals within a population can restrict the remaining population's access to that resource. Efficiency is determined by the amount of energy expended in converting a resource, it does not matter whether that energy is expended in the acquisition of the resource or in the conversion of the resource. The most efficient resources are those that require the individual to expend the least energy to acquire and convert.

Environments that induce development, such as the embryonic environment of a fertilized egg cell or an evolutionary period in the biosystem, have abundant and diverse resources and are sparsely populated. When a seed population initially enters a developmental environment, it will first consume the most efficient resources available, which are those that require the least energy to access and to convert. In a resource dense and diverse environment, the most efficient resources may constitute only a portion of the diversity of the resources available in the environment. The limiting resource in the set of most efficient resources will determine the population capacity that can be supported by the set of most efficient resources. When the flow of population reaches the capacity of the limiting most efficient resource, additional individuals must either compete for that resource, substitute a less efficient resource, or make do without it. This alters the mix of resources in the environment of those additional individuals from that of the original population.

The use of a resource to its capacity by a segment of the population establishes priority and prevents other members of the population from utilizing that resource. Any competition for a resource expends energy which increases the energy necessary to access that resource, decreasing its efficiency and increasing the relative efficiency of otherwise less efficient resources that are readily available and can be substituted. Individuals, therefore, tend not to compete for resources when alternative resources are available.

The expansion of a population to the capacity of the limiting resource, in the set of most efficient resources, effectively decreases the efficiency of the limiting resource to additional members of the expanding population, while increasing the efficiency of any resource that can be substituted for the limiting resource. Any growth of the population beyond the capacity of the limiting most efficient resource must either utilize a substitute resource or function without that resource. This resource difference between the two segments of the population divides their environment and segregates the additional population from the original population along the distribution of the limiting most efficient resource.

As a population expands into an environment, it will distribute itself where resources are most easily and efficiently converted. Once those resources are utilized to capacity, the population will then expand into regions of the environment with less efficient resources.

It is the expanding population that pushes individuals into successively less efficient regions of the environment. As these regions are filled, their utilization at capacity prevents movement between regions, causing the initial population to be divided into multiple segregated regions of greater and lesser resource efficiency with little population movement between regions. This causes groups of individuals within a population to self-segregate based on the relative efficiency of various resources.

Suppose an environment with only two resources **A** and **B**, where **B** is a substitute resource for resource **A**, and **B** is also a less efficient resource than resource **A**. As an initial population expands to the capacity of its environment it will preferentially utilize resource **A** because it is the more efficient resource. Once resource **A** is utilized to capacity, any increase to the population will be segregated to using resource **B**. This divides the population between resource **A** and **B** and divides the environment into environments **A** and **B**. The population of environment **B** will then expand to the capacity of resource **B**. The division of the population into environments **A** and **B** initiates developmental divergence in subsequent generations of each population.

The division of a population by resource use does not necessarily segregate the population spatially. Both less and more efficient resources may be distributed throughout the same geographical area. Differences between individuals in their priority in access to resources is enough to sufficiently segregate a population. The cichlid species of Lake Victoria and Lake Malawi are examples of diversification occurring within segregated populations cohabiting the same geographical area.

A population is delineated by its environment, and self-segregation divides both the population and the environment. Once a group is segregated from the initial population, it is no longer a single population, but instead a single species, or type, divided into separate populations, each with its own environment distinguished by the efficient resource that initiated the segregation of the population. This division of a population into separate environments initiates developmental divergence between the populations and differentiation in their forms.

While differentiation is triggered by a population reaching the capacity of its scarcest most efficient resource, it is not the scarcest most efficient resource that causes the population to differentiate. Rather it is the changed environment caused by the loss of access to that resource, and the increasingly efficient conversion of the changed environment by subsequent generations of the population segregated to it. It is the change in the resources of the environment that directs development in the segregated population toward a modified form.

Form is modified though development by altering the distribution of units of form, and developmental environments produced from the conversion of different resource sets do not necessarily organize compositional units of form in the same way. The previous chapter provides several examples of altered environments at different levels of organization modifying the development of form.

As an individual develops, it's units of form move efficiently which conforms the individual's development to its position. In the process of modifying a form to a changed environment, each subsequent generation of a lineage starts where its preceding generation left off, such that the series of developmental processes that occur throughout the successive generations of a lineage modify the form of the individual to a position in its new environment. Differentiation between the initial population and the segregated portion of the population is caused by subsequent generations of the segregated population becoming more efficient at converting the less efficient resources to which they are segregated.

Units of form moving efficiently within a developing system will organize the system's structure toward a form that more efficiently converts the resources of its position. Because units of form move efficiently along the path of least resistance, any variation in their pattern of movement is toward greater efficiency. This does not mean that the system produced by that unit of form's change in motion will necessarily be more efficient, but over generations of development, feedback loops will guide the efficient movement of units of form toward an organization that produces a system more efficient at converting the resources of its position. Over several generations, both the efficient movement of units of form during development and the feedback loops between levels of organization will cause each successive generation of a form to conform more efficiently to its position and environment. To whatever extent efficiency can be increased in a diverging population, greater differentiation and adaptation may ensue. The modification of form in this manner will continue over successive generations until no greater efficiency can be achieved.

To see the process of differentiation in action within the biosystem we need only to find an incipient population expanding into a diverse environment, and here invasive species provide the example. Over the past few decades both genetic and field research have shown that invasive species demonstrate both the genetic and phenotypic flexibility that indicates speciation in progress. In classical evolutionary theory, genetic mutation is random, yet polyploidy – genetic doubling – occurs with much greater frequency among invasive plants than among angiosperms in general. The invasive Australian cane toad is an example that is showing signs of phenotypic divergence at the leading edges of its invasion.

Differentiation increases the efficiency of the individual by specializing its form to convert less efficient resources more efficiently. The differentiation process of population expansion, segregation, and modification can be repeated until there is no more capacity to expand, or too little resource variation to segregate the population. However, if repeated over time and space, it can eventually lead to adaptation and speciation. If differentiation achieves adaptation, the population may be able to further expand its environment to include resources that it was unable to access or utilize prior to having the adaptation, thus increasing the range and diversity of its environment which can lead to further differentiation.

The continued modification of a form toward greater efficiency can, over several generations, lead to adaptation. A population diversifies by producing different adaptions in different lineages of its descendants. Diversification may occur when a population splits into multiple segregated populations that each adapt to a different environment, or through a succession of splits in which the splitting populations adapts and expands to its new environment then splits again, producing a new form with each split.

While most of the resources used by the populations that form a system are common to all the populations in that system, there are some that are specific in type or volume to specific populations. These are the resources that differentiate the environments within a system. An environment that induces a seed population to grow and diversify into a system must not only provide a volume of resources that will support the population's rapid expansion, but also a variety of resources sufficient to induce its diversification.

Resource diversity is initially high in a developmental environment, and it can be increased. The functioning of a newly differentiated population can modify a developmental environment by adding new resources to it, which may enable further diversification in other populations; or the efficiency gained by differentiating may be used to convert additional resources that were unusable prior to differentiation. But while differentiation can increase the variety of useful resources in one's environment, and in some instances add new resources to the environments of other populations, specialization through differentiation can also cause a population to lose the ability to convert resources it was previously able to convert.

Over the course of a system's development, diversification increasingly specializes a system's populations while at the same time subdividing its environments. Environments become increasingly less resource diverse and the positions within them more homogenous. An environment of a mature population is relatively homogenous throughout its range as compared to a developmental environment, and although positions can vary, the variety, availability, and use of resources is similar throughout the positions of a mature population's environment. The less diverse an environment, the fewer alternative resources there are available, and the less opportunity there is to achieve greater efficiency. Mature populations do not diversify because their environments do not contain the resource diversity sufficient to induce further diversification. Diversification ends when no greater efficiency can be achieved.

SEQUENTIAL DEVELOPMENT

Embryonic development is a cladogenetic process in which a seed population, usually containing one fertilized individual, expands into a large, unoccupied and resource diverse environment. By expanding and splitting the embryonic environment, the seed population diversifies into the integrated populations that form the system. In all organisms, embryonic development reproduces the same form with an extremely high degree of accuracy. It is as accurate as the sand in an hourglass producing the same form in the bottom globe each time it is flipped. To produce this accuracy, the sequence and timing of differentiation must by tightly controlled.

The sequence of diversification is guided by the relative efficiency of the resources in its environment. As an initial population reaches the capacity of its scarcest most efficient resource, it diversifies utilizing the next most efficient resource. All subsequent populations diversify by substituting increasingly less efficient resources as the capacity of their scarcest most efficient resource is met. The course of diversification is then set by the distribution and relative efficiency of the resources contained within the developmental environment, both in its initial state and as it evolves during development.

A population develops in a particular environment because its individuals have the functional ability to convert those resources. However, environments tend to provide a variety of resources which are not distributed uniformly, have different purposes, and, when interchangeable, are not equally convertible. Additionally, the topology of an environment and the physical structures within it can affect the distribution of environmental resources over both space and time. These things affect the circulation of the population and the flow of environmental resources to the population, which in turn can affect the sequence of development. This is especially true in the molecular environments of a cell where the molecular resources in these environments can be segregated and compartmentalized by such barriers as diffusion gradients, temperature, chemical affinities, as well as the barriers created by the physical structure of the cell.

Development in a structured environment follows a particular pathway not because the individual is programmed to take a specific action at a particular time and place but rather because the conditions in a structured system direct individual motion by either limiting, blocking, or enhancing alternative pathways. A population merely needs to move efficiently, and similar populations in similar environments have the same efficiencies. This movement is directed by the distribution of their environmental resources, the diversity and volume of those resources, and the physical structure in which the environment exists. Just as the walls, doors, and furniture in a house direct how one moves through it, the physical structure in which an environment resides affects how resources and individuals circulate through an environment. The distribution of resources directs the circulation and flow of units of form. Because most of the course of development is produced by the movement of a great many individuals reacting to their environment, rather than one or a few individuals reacting to a position, an initial developmental position or environment, does not need to be exactly structured. Repetitive development will require the correct diversity of resources but not necessarily exact volumes or distributions, relative volumes and distributions should suffice. An hourglass for example can produce a similar form in its lower globe with either more or less sand.

Individuals preferentially utilize those resources that are easiest to access and convert, and within a cell, individual molecules react preferentially with the most accessible molecules and with those that form the strongest molecular bonds. These differences in access and convertibility (or reactivity in molecular populations) between individuals and resources creates a probability ranking in which each action in a developmental process has a greater or lesser probability of occurring, given the present circumstances. The sequence in which a population diversifies, and the variety to which it differentiates, is then determined by the relative efficiency of the scarcest resources.

Differentiation is forced to occur by the expanding population, which when reaching the capacity of the scarcest most efficient resource, must utilize a substitute resource, function without the resource, or cease expanding. The capacity of an environment and the scarcest resource in that environment determines the timing of diversification, while the existence of substitute resources and their relative efficiencies determine the sequence of diversification and direct the variety that diversification takes.

To illustrate this, suppose for example that we have a series of buckets with each bucket in the series being slightly shorter than the previous bucket, and the buckets are aligned so that, when full, the spout of each bucket will pour water into the next shorter bucket in the line. We also have a spigot that pours water into the first and tallest bucket. When the spigot is opened the buckets will fill in sequence starting with the first bucket. When the first bucket fills to the height its spout, water will then flow into the second bucket, filling it, and this will continue filling each bucket in sequence. In this example the buckets represent a developmental environment, the water represents the population filling that environment, and the spout on each bucket represents the scarcest resource. The sequential arrangement of the spouts from highest to lowest represents the order of relative efficiency of the scarcest resources, and the water in each bucket represents a differentiating population.

Assuming the water always flows at the same rate, the timing and sequence in which the buckets fill will be the same each time the spigot is opened. The water in this example, which represents a segregated population in each bucket, does nothing more than move efficiently by following the path of least resistance. We should expect no less from a seed population in a developmental environment. The sequence and timing of diversification is directed by the relative efficiency and capacity of the scarcest most efficient resources, and because a population moves efficiently, its flow will deliver individuals where needed in the proper sequence and time. So, to the extent that a seed population and its developmental environment are both identically reproduced, their development process should be repeated.

Protein Folding Sequence
Protein folding appears to be a process analogous to diversification in that the sequence of folding is determined by efficient movement along the path of least resistance, or stated differently, along the path of least energy expended.

A protein is synthesized initially as a linear polypeptide chain extruded from the ribosome. Each protein type has a specific order of amino acids in its polypeptide chain that specifies the final folded form of the protein, and it is this final folded form that makes the protein functional.

Protein folding occurs almost instantaneously and in two stages. In the first stage different regions of the polypeptide chain fold independently to produce the secondary structure of the protein, the initial linear polypeptide chain being its primary structure. In the second stage, the folded areas of the polypeptide chain fold together to produce the protein's final, tertiary form. Protein folding is produced by the formation of peptide bonds between amino acid residues along the polypeptide chain.

It is thought that the final form of the protein is the lowest energy state of the initial polypeptide chain, and the sequence of folding is determined by the free energy in the initial chain. If this is correct, we may then assume that the formation of peptide bonds between linearly arranged amino acid residues to be sequentially efficient – just as cell differentiation in embryogenesis is sequentially efficient – with those bonds requiring the least energy being the first bonds formed.

The bonding efficiency of two residues may depend on a variety of factors such as the bonding propensity of the two residues, the distance apart, their orientation to each other, and the aqueous medium in which the bonding occurs. These factors fall into two categories: the order of amino acids in the initial polypeptide chain and the content of the aqueous medium. If both remain consistent, then identical polypeptide chains should always fold into identical proteins. However, because the path of least resistance for each potential peptide bond changes as each peptide bond forms, the process of protein folding appears to be a typical "N-body" problem which would make it difficult to use a protein's primary structure to deduce its tertiary structure, especially as the length of the initial polypeptide chain increases.

KARYOTYPE EVOLUTION

It can be assumed that in multicellular organisms the karyotype of the genome evolves along with the phenotype of the individual. This is because there is no known example in which two distinct species share the same karyotype. Furthermore, both chromosomal polymorphism as well as polyploidy tend to be associated with species that have recently undergone, or may be in the process of undergoing, divergence. Hence, flux in the genotype appears to coincide with flux in the phenotype. This seems to suggest that genetic evolution, because it coincides with phenotypic evolution, is the cause of phenotypic evolution, but this is not necessarily the case in multicellular organisms.

Most evolutionary changes in the phenotype of multicellular organisms are accomplished through the reconfiguration of cellular distributions, rather than through novel genetic mutations. Both cell types and functional gene types remain quite similar throughout the species represented within a taxonomic family, and novel genes specific to a species make up only a very small percentage of the genome. Moreover, functional genes tend to be conserved even in very distantly related organisms. What then is the connection between genetic flux, such as polyploidy, and phenotypic divergence?

The expression of a gene is controlled by both the transcription of gene and the number of times the gene is translated by the ribosome, so one possibility is that gene duplication may allow greater expression of a gene. The increased expression of a gene, or genes, does not necessarily change the type of cell, especially within taxonomic families, but may alter the functional efficiency of the cell type. This can alter the distribution of the cell populations during the formation of the multicellular organism, which will alter the phenotype of the organism. Because cell distribution in the multicellular form is determined by the cell's environment, the genetic flux caused by polyploidy is not then directing phenotypic changes in multicellular organization but rather allowing them to occur.

The probable causal pathway in which the maternal phenotype affects the evolution of its offspring's phenotype is through the maternal cytoplasm by inducing polyploidy. Polyploidy allows variation in genetic expression, which may alter the functional efficiency of the cell lineages during embryonic development. Altered efficiencies in the cell populations of a developing organism affects the distribution of those populations which, in turn, alter the mature form of the organism. How such efficiencies are retained and incorporated into the development of future generations is unknown, but the avenue to do so does exist, and the result appears to be a karyotype that evolves along with the evolution of the species.

STASIS

While stasis is commonly defined as the lack of adaptive morphological change in an organism over some period, in the structural perspective stasis may be defined as a period during which the efficiency of a biological system does not increase. In contrast, diversification increases the efficiency of a system, while morphological adaptation (differentiation) increases the efficiency of the organism in its environment. Both evolution and development are periods of increasing efficiency. More generally, stasis may be defined as a stable and repetitive pattern of biological motion that produces the same form each time it cycles.

Every biological form is the pattern of motion of its units of form. For example, a cell is the pattern in which its molecules move, and an organism is the pattern in which its cells move. Patterns of motion repeat themselves and combine with one another to form longer patterns. Evolution is one long pattern of motion of the organismal units of form that make up the biosystem. The Krebs cycle is one of the shortest pattens of motion in biology, and one that is universal to all life. Embryonic development is the pattern of motion of cells within the lifecycle of an organism, and the organism's lifecycle is an extended pattern of motion of its cells. Each organism's lifecycle is a repeat of its parent's lifecycle, and the individuals of a population are all repeating a similar lifecycle.

The maturity of a system produces stable patterns of motion in its units of form, while development produces diverging patterns in its units of form. This suggests that stasis is a consequence of maturity while variation is consequence of development, and this is essentially correct. However, a developmental process normally follows a very repetitive pattern to consistently produce the same form each time it cycles, suggesting that a pattern of development may also be in stasis. This dilemma is resolved simply by considering the frame of reference in which the pattern of development is taking place. For example, if we consider the development of a multicellular organism then its pattern of development is produced by the continually diverging development of its cells. The cell populations in the developing organism are not in stasis. However, if the development of the same organism is considered from the perspective of the greater system in which the organism resides then we find that the pattern of cell divergence occurring in the development of the organism is consistent throughout the population of organisms. From this perspective the pattern of divergence is a stable pattern of motion that is repeated in the development of each organism in the population.

The frame of reference for determining stasis is the level of organization in which the pattern of motion is being considered. Developmental patterns of motion when considered individually are never in stasis but when considered from the perspective of the system in which they take place then they tend to remain stable and repetitive for extended periods.

The developmental pattern of motion that produces an organism becomes repetitive when similar individuals develop in similar positions. Since the development of an individual conforms to its position, adaptive variations in development can be attributed variations in position. The more stable and uniform the positions are throughout and environment, the more uniform will be the development that occurs throughout its positions. Positions remain stable to the extent that their environment remains stable, and an environment is stable to the extent that its system is stable. This means that to the extent that there is stability in greater levels organization, the development that takes place within sublevels of organization will be repetitive. Systems, and the environments produced within them, are most stable when they are mature.

Environments in mature systems are relatively stable in comparison to environments within developing systems because of the homeostatic equilibrium that a mature system maintains. They are also less resource diverse due to the environmental division that takes place during the mature system's development, and the positions within them are smaller and more confining due to their populations being at capacity. These factors cause the positions throughout an environment within a mature system to provide a more uniform set of resources than in a developing system. The similarity of positions within the environments of a mature system means that the individuals developing in those positions will develop along similar developmental pathways. Repetitive patterns of development then arise within mature systems because they maintain stable internal environments that sustain uniform positions. Thus, for example, when the cells in our body expire, they are replaced with cells that developed in positions that are the same as those of the cells they replaced. The stasis in form that appears in the fossil record is a result of the biosystem maintaining stable environments that sustain stable positions in which that form is produced.

Biological form is a pattern of motion, and patterns of motion are stable when they become repetitive. The more repetitive a pattern of motion, the more stable the form produced by that pattern. Stasis is a relative measure of the duration that a pattern of motion is repeated, and it is simply a characteristic expressed by units of form when they inhabit mature systems, while morphological variation is a characteristic they exhibit during a system's development.

Adaptation is always toward greater efficiency in converting accessible resources, so stasis begins when a developing system and the units of form within it can no longer increase their efficiency. Hence, an environment does not need to be perfectly uniform or perfectly stable, rather the variations that occur in a population simply need to be insufficient to enable the individuals inhabiting the environment to increase their efficiency.

Mature systems, by producing stable environments containing generally uniform positions and confining the efficient movement of individuals to those positions, produce repetitive patterns of development in their units of form. Repetitiveness is caused by the efficient movement of similar individuals in similar environments.

Given that resources have different relative efficiencies which direct the course of units of form as they move along the path of least resistance, individuals with the same functional abilities placed in similarly structured developmental positions containing the same diversity of resources, should move in the same pattern causing their pattern of development to produce the same form. Conversely, variations in the structure and resource diversity of a position will cause variations in development and in the form produced. The emergence of biological form is fundamentally a result of the space contained within an environment confining the efficient movement of the units of form that reside within it.

AFTERWORD

In advancing a new philosophy of biology it is important to explain how that new philosophy improves upon the old. While I believe each of the many concepts presented in this book adds to the understanding of evolution, the overarching achievement of structural theory beyond the Darwinian paradigm is that variation is no longer random but instead variation is directed toward adaptive form. This change allows a causal explanation, not found in Darwinian theory, for the pattern of evolution apparent in the fossil record.

The central tenet of Darwinian theory is that natural selection selects for the genotype through the selection of the phenotype. Natural selection, as it was conceived, selects from among individuals based on the variations their individual phenotypes may possess, but it states nothing about how individuals come to possess their variations, other than that they are inherited. Neo-Darwinian theory by presupposing that heritable factors must control development designated the genome, passed from parent to offspring, as the structure that both transmits inheritance and controls development. Variations in form must then arise from the differential assortment of genes that occurs during reproduction. Thus, each new individual is a new variation.

In this reasoning, variation occurs constantly but the phenotypic product of that genetic variation is not necessarily directed toward an adaptive phenotype. Every variation has an equal probability of being to any degree advantageous, disadvantageous or neutral. The probability that an advantageous variation occurs is considered by Darwinian theory to be random.

Random in the Darwinian interpretation means that nothing induces the production of an advantageous variation. Rather advantageous variations, like maladaptive and neutral variations, are simply a happenstance of genetic crossing. Advantageous variations are then selected from among all variations through natural selection.

According to Darwin, all populations increase at a geometrical rate. From this inevitably follows a struggle for existence that will select and preserve those individuals having variations to any degree favorable. He explained it in this way:

"Hence, we may confidently assert that all plants and animals are tending to increase at a geometrical ratio… There is no exception to the rule…as more individuals are produced than can possibly survive, there must in every case be a struggle for existence…"

Here Darwin asserts that all populations produce more individuals than can survive, and this in every case causes a struggle for existence. If over reproduction is a continual condition for all life and it is the cause of selection, then selection must be an ever-present force driving evolution, and favorable variations should always be selected whenever they arise.

Adaptations are the accumulation of advantageous variations, and they represent increments in the process of evolution. If variation is truly random and selection is continual, then the rate of adaptation, in other words the pace of evolution, should correspond to the number of attempts at variation. An attempt at variation is simply the production of an individual, as each individual is a single variation. The pace of evolution should then reflect the number of attempts at variation, in other words, the more individuals produced, the more variations that occur and the greater the probability of accumulating enough variations to produce an adaptation.

If variation is random and selection is continual, then the patten of evolution as evidenced in the fossil record should be randomly distributed across taxa, in degree, and throughout time. Any variance in this random distribution should be associated with a greater or lesser production of individuals. More or fewer individuals produced, increases or decreases, respectively, the probability that advantageous variations occur. Such a distribution is clearly not evident in the fossil record. In fact, the fossil record exhibits two distinct patterns of evolution, cladogenesis and anagenesis, neither of which expresses a random distribution.

The greatest amount of evolutionary change occurs through the proliferation of species during cladogenesis, and yet periods of cladogenesis are often condensed in time into the thin layers between geological strata. Although short in duration, these periods often display the concurrent evolution of unrelated taxa, which suggests that evolutionary variation is neither random nor genetically induced, but rather induced by the conditions of the period.

The phyletic evolution of anagenesis is quite the opposite. Its occurrence is thinly distributed over both time and taxa during very long geological periods of stasis, and it rarely produces much evolutionary change beyond the species or subspecies.

If variation is random, the continual force of selection should not produce two distinctive patterns of evolution. By assuming selection to be continual and variation to be random, Darwinian theory restricts the pace of evolution to being determined by only the probability at which advantageous variations occur, which is some percentage of the number of individuals produced. This prohibits any factor, other than the number of individuals produced, from affecting the pace at which evolution proceeds. Hence, while paleontologists have long recognized distinct differences between phyletic and speciational patterns of evolution, the philosophy of Darwinian theory contains no mechanism with which to understand why those differences should occur. This problem often debated in the context of "gradualism" has existed ever since the inception of Darwinian theory and has never been coherently rectified.

Structural theory is quite different. It is a theory of motion rather than of inheritance, and this relieves it from the strictures of genetic determinism. In structural theory variations arise not from a difference in units of form but from a difference in the distribution of units of form. Units of form are distributed by developmental processes, so it is a difference or change in a developmental process that produces a variation in form. In other words, a genome has multiple developmental potentials, and it is the progression of the developmental process that determines which variations are realized. The genome is only one of several factors that influences development.

Development is a pattern of motion produced by the interaction between a population and its environment. Environments are produced at multiple levels of organization within the developing structure of an organism, and it is these environments at each level of organization that direct development. An environment through its distribution confines and directs the efficient movement of its population. This causes the distribution of a population to conform to the distribution of its environment, and it is the distribution of the population that gives shape to phenotypic form. Variations arise when patterns of development are altered by changing environments.

Individuals move efficiently and the distribution of their environment determines the patten of movement that is most efficient. While environment always directs variation in movement toward an efficient pattern, the multilevel organization of biological form means that efficient movement at one level of organization does not necessarily induce more efficient movement at a greater level of organization. However, the feedback loops between levels of organization along with the homeostatic movement of populations within levels of organization cause the repeated development of an organism to always adjust the distribution of its compositional populations toward a form that efficiently converts the resources of its environment. Thus, variation is always adaptive and will over repeated generations direct evolution toward an efficient organismal form.

The primary philosophical improvement of structural theory is that it establishes a determinative cause for adaptive variation. It recognizes form as a patten of motion and that motion is efficient which causes any variation in motion to always be toward greater efficiency and thus adaptive.

Moreover, by identifying variation as a change in distribution and recognizing the multiple levels at which biological structure is organized, it can describe the physical process through which adaptive variations, to the exclusion of maladaptive and neutral variations, are induced and directed toward the production of adaptive form. This allows the concepts of structural theory to explain not only cladogenesis and anagenesis but also morphogenesis.

In contrast Darwinian theory assumes that variations arise through the genetic reassortment that occurs in sexual reproduction or through replication errors that happen in asexual reproduction. However, whether a variation is advantageous, disadvantageous, or neutral has no determinative cause in Darwinian theory. For all its humanistic and secular connotations, Darwinian theory remains reliant on fortuitous happenstance - which some have claimed as divine intervention – to produce the series of adaptive variations necessary to move evolution forward. This leaves Darwinian theory wholly inadequate to explain the patterns of evolution expressed in the fossil record, much less morphogenesis.

GLOSSARY

Adaptation
A change in the distribution of a population which causes a change in the form of its system that is sufficient to cause a change in the functioning of the system. A change in form that engenders a change in function. A functional change in form.

Anagenesis
A change in form throughout a population caused by a change in its environment. A phyletic transformation within a single population.

Biological Structure
The organization of the units of structure that compose a biological form.

Biological Reaction
The reaction between a population and its environment that produces more population.

Biosystem
A singular system that is composed of all the biological organisms that exist in the biosphere.

Cladogenesis
The diversification of lineage into multiple forms caused by an expanding population splitting its environment.

Complexity
Relative motion in multiple frames of reference. The multiple tiered frames of reference in which the organized movement of units of form takes place.

Conversion or Resource Conversion
The modification of resources. The transformation of resources through metabolism and reproduction into population.

Circulation
The movement of individuals within an environment.

Clade
A group of diverse species evolved from a common lineage.

Development
The formation of a system. The changing distribution of the populations forming a system.

Differentiation
The splitting of a population through the division of its environment to form a new type or species. A variation in a pattern of motion that produces a form different than that from which it originated.

Direct Resources
Those resources utilized directly by a population to support its subsistence and reproduction. Resources that need no further conversion to be utilized by the population.

Distribution
The number and location of the individuals in a population. A population's spatial organization.

Diversification
The splitting of a common ancestor into the multiple forms. The process of speciation or differentiation.

Ecosystem
A feature of the biosystem. An intermediate organization of populations within the biosystem. A homeostatic and interdependent group of mostly invasive species.

Efficiency
The property of motion that causes all movement to be in the direction of least resistance given the immediate circumstances. The property of motion that gives direction to movement. Expending the least energy for the greatest motion. The ability to retain energy within a system.

Environment
The direct resources that a population can access. The area in which a population moves. A set of resources that confines the movement of a population.

Environmental Resources
The direct resources of a population.

Evolution
A change in the distribution of a population which causes a change in the form of its system that is sufficient to engender a change in the functioning of the system. The diversification of form through adaptation.

Feature
A distinguishable part of a larger entity, which may or may not have a specific purpose or function, but which cannot be delineated from the whole because its compositional parts move with the larger entity.

Flow
The movement of individuals into and out of a population through reproduction and mortality.

Form
The spatial distribution of the compositional parts of a structure. The distribution of a system's units of form.

Function
The ability of an individual to convert resource. An individual's ability and manner of movement.

Homeostasis
The constant adjustment of a system toward equilibrium with its position.

Homeostatic Effect
The constraint on growth, dispersal, and movement (circulation and flow) that the populations within a system have on each other.

Indirect Resources
All the processes and materials necessary to produce direct resources. Those resources from which direct resources are produced.

Individual
A reproducing or reproducible biological structure that occupies a position within its population and environment. A unit of biological form.

Level of Organization
A frame of reference for biological organization, identified by the complexity of the unit of form being organized and the complexity of the system produced.

Maturity
The point at which the consumption of resources by a system reaches equilibrium with the capacity of the resources provided by its position. The period during which a system maintains equilibrium with the capacity of its position. A mature population is one that is at equilibrium with the capacity its environment.

Metabolism
The conversion of resources into population through catabolism and anabolism.

Modification
A change in form that does not necessarily rise to the level of an adaptation. A variation.

Motion, Movement
A change in spatial organization.

Organism
A reproducing or reproducible biological system.

Organization
The movement of objects relative to one another within the same frame of reference.

Pattern of Motion
A sequence of movement through a series of locations arranged in space.

Population
A set of functionally and phenotypically similar individuals occupying the same environment.

Position
The volume and variety of direct resources available to an individual or system. That portion of a population's environment that is available to a specific individual. The area in which an individual moves and accesses resources.

Reproduction
The conversion of environmental resources into population.

Replication
Reproduction.

Stability
A relative measure of the duration that a pattern of motion is repeated.

Stasis
The continued repetition of a pattern of motion. A period in which the efficiency of a biological system does not increase. A state of dynamic equilibrium such as maturity.

Structure
The compositional elements of a form. The distribution of individuals or populations that compose a system.

Structural Unit
A unit of form.

System
A group of diverse and interdependent populations that acts as an individual and occupy a position. A system originates, grows, and dies as an individual, and is usually able to reproduce.

Trophic System
A biological system in which energy and biomass are converted from one form into another.

Unit of Form
An individual. A unit of structure within a biological system or population.

Unit of Structure
A unit of form

Variation
A difference in the distribution of a population that alters the form of a system, but which is not sufficient to change the functioning of the system. A difference in form that does not produce a change in function. Dynamically: A change in the pattern of movement of a population.

Variety
A set of individuals that have the same form and function. A species or subspecies.

NOTES

NOTES

www.ingramcontent.com/pod-product-compliance
Lightning Source LLC
Chambersburg PA
CBHW072017230526
45479CB00008B/69